ICMI Study Series Editors A.G. Howson and J.-P. Kahane

The Popularization of Mathematics

Edited by
A.G. Howson
University of Southampton
and J.-P. Kahane
Université de Paris-Sud

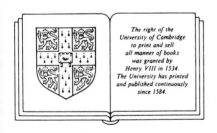

CAMBRIDGE UNIVERSITY PRESS

Cambridge

New York   Port Chester   Melbourne   Sydney

Published by the Press Syndicate of the University of Cambridge
The Pitt Building, Trumpington Street, Cambridge CB2 1RP
40 West 20th Street, New York, NY 10011, USA
10, Stamford Road, Oakleigh, Melbourne 3166, Australia
© Cambridge University Press 1990

First published 1990

Printed in Great Britain at the University Press, Cambridge

Library of Congress Cataloguing in Publication data available

British Library Cataloguing in Publication data available

ISBN   0-521-40319-7   Hardback ✓
       0-521-40867-9   Paperback

Contents

FOREWORD

A Study Overview   1
    Geoffrey Howson and Jean-Pierre Kahane

Mathematics in Different Cultures   38
    Report of the Working Group

Mathematics for the Public   41
    Edward J. Barbeau

Making a Mathematical Exhibition   51
    Ronnie Brown and Tim Porter

The Role of Mathematical Competitions in the Popularization of Mathematics in Czechoslovakia   65
    Vladimir Burjan and Antonin Vrba

Games and Mathematics   79
    Miguel de Guzmân

Mathematics and the Media   89
    Michele Emmer

Square One TV : A Venture in the Popularization of Mathematics   103
    Edward Esty and Joel Schneider

Contents

Frogs and Candles - Tales from a Mathematics Workshop *Gillian Hatch and Christine Shiu*	112
Mathematics in Prime-Time Television: The Story of Fun and Games *Celia Hoyles*	124
Cultural Alienation and Mathematics *Gordon Knight*	136
Solving the Problem of Popularizing Mathematics Through Problems *Mogens Larsen*	144
Popularizing Mathematics at the Undergraduate Level *Brian Mortimer and John Poland*	151
The Popularization of Mathematics in Hungary *Tibor Nemetz*	160
Sowing Mathematical Seeds in the Local Professional Community *Tony Shannon*	170
Mathematical News that's Fit to Print *Lynn Steen*	176
Christmas Lectures and Mathematics Masterclasses *Christopher Zeeman*	194
Some Aspects of the Popularization of Mathematics in China *D.Z. Zhang, H.K. Liu and S. Yu*	207

Foreword

A.G. HOWSON AND J.-P.KAHANE

A STUDY AND AN EVENT

This book provides a description of the fifth in the series of ICMI studies. Earlier studies have considered the themes: the influence of computers and informatics on mathematics and its teaching, school mathematics in the 1990s, mathematics as a service subject, and mathematics and cognition. The studies have varied in structure, but most have followed the same general pattern. As a first step, a programme committee appointed by ICMI has met, surveyed the problem area, and produced a discussion document. These documents have sought to identify major questions or challenges to which mathematics educators need to make responses and have invited readers to submit papers describing their reactions and experiences. These papers, along with plenary presentations from specially invited speakers, have then formed the basis for the work of an international symposium.

In the case of this study on the popularization of mathematics, the discussion document, written by A G Howson, J-P Kahane and H Pollak, was first published in *L'Enseignement Mathématique*. It later appeared in *The Bulletin of the IMA*, *The Notices of the AMS*, and in translation.

The response to this paper was most encouraging. Many reactions were submitted to us: so much so that, to our regret, not all those who contributed papers could be invited to participate in the seminar held at Leeds, England, from 17-22 September, 1989. Elsewhere the discussion document provided a basis for local meetings. Thus, for example, the Spanish Federation of Societies of Teachers of Mathematics organised a meeting in Granada in June 1989 and produced their own national document, *Hacia unas matemáticas populares*. Clearly the theme of popularization was one which appealed to, and concerned, mathematicians and mathematics educators throughout the world. Some 80 such persons from 20 different countries were able to accept the invitation to attend the Leeds meeting.

The symposium commenced with three plenary talks, by Christopher Zeeman, Alain Connes and Lynn Steen, before it turned to the consideration of the problems of popularization through particular media. Plenary discussions on the different media were introduced by a variety of conference members drawn from various countries, some being mathematicians, some mathematics educators, and others, such as Roger Lesgards, President of la Cité des Sciences et de l'Industrie, Paris, coming from outside mathematics. The work of the symposium was also carried out in ten or so discussion groups each devoted to a specific theme, for example, TV and films, radio, games and puzzles, competitions, the image of mathematics and mathematicians, and mathematics in different cultures. At the closing plenary session, Henry Pollak presented a personal overview of the meeting.

This, then, is a brief survey of the symposium, its planning and programme. In the papers which follow we shall indicate in more detail the outcomes of our talks and discussions. It must not be thought that these yielded immediate, internationally acceptable answers to the problems which we sought to identify. Yet they did provide new insights, for example, the emergence of an entirely new target audience for popularisation, the vast number of retired people eager to find intellectual challenges and stimulation; they allowed experiences to be exchanged and compared, and in certain instances, such as TV, they revealed that the time was now ripe for a much more detailed investigation of the ways in which a particular medium is being, and might be, used for the purposes of teaching and popularizing mathematics.

What this foreword has so far omitted entirely to describe is the 'event' which accompanied the Leeds meeting.

The study arose from a desire more clearly to identify the 'why, to whom, how and what?' of popularization. The danger of such an approach is that it might lead one into theoretical abstractions and away from the realities of life. One way to prevent this was by ensuring that practitioners, TV producers, film makers, authors, etc, were well represented at the symposium. Another was to link the meeting with a large-scale attempt at popularization: this we were able to do.

At an early stage of the planning for the meeting, we were able to involve David Crighton, Chairman of the Joint Mathematical Council of the UK, and Geoffrey Wain, a mathematics educator at Leeds University, in our work. They, with the backing of the Joint Mathematical Council and the Royal Society, soon set about the task of planning what was to become the greatest event aimed at popularizing mathematics ever to be mounted in England: indeed it may well have had no parallels elsewhere.

Together with a specially-appointed Planning Committee, representative of all aspects of British mathematics and mathematical education and serviced with great efficiency and enthusiasm by Jill Nelson, the Royal Society's Education Officer, they prepared a 'Pop Maths Roadshow' of outstanding variety and richness.

The centrepiece of the roadshow was an exhibition covering over 2000 square metres. Here, nearly thirty distinct exhibits could be found including, for instance, the Frontiers of Chaos collection from Bremen (together, of course, with micros on which visitors could generate their own pictures), an English version of Horizons Mathématiques, a touring exhibition from M. Lesgards' museum which provided many opportunities for 'hands-on' activities, Common Threads, an intriguing collection of textiles from around the world illustrating the place that mathematics plays in their design and construction, exhibitions from Lille and Dortmund, a collection of mathematical games assembled from many countries in Africa and elsewhere, and an exceptionally beautiful and impressive collection of sculptures and tapestries by the artist John Robinson which used the theory of knots for their inspiration and which complemented an exhibition on knots designed in Bangor, Wales and described more fully later in this book.

Although no visitor to the Roadshow would have failed to be impressed by this international display, it is likely that an adult visitor would have been equally taken by the sight of so many schoolchildren, from the age of seven upwards, eagerly participating in impromptu activities, exploring the tile maze, or trying out some of the large collection of games and puzzles which had been assembled. Elsewhere a group of 13-14 year-olds worked at a variety of mathematical problems and were keen to discuss their work with visitors. Should the visitor have felt the need to sit down for a time, then he or she could have visited the films and videos on mathematical themes which ran continuously throughout the roadshow. Here were collected a rich selection drawn from several countries, including work by Michele Emmer (Italy), the beautiful film on Ramanujan made for England's Channel Four, and selections from the entertaining and provocative output of the New York Children's Television Workshop.

On most days three popular lectures were offered; two in the afternoon, and intended primarily for schoolchildren, and one in the evening. The menu was a rich one; ranging from 'funtastic' mathematics, a presentation by the TV personality, Johnny Ball, to a repeat of the lecture which Christopher Zeeman had presented at the Royal Society to mark the presentation to him of the Faraday Medal for popularizing mathematics and science; from a talk on the combinatorics of juggling, complete with a demonstration whilst riding a unicycle, to one on the theory of knots given by Oxford University's youngest D. Phil., the seventeen year-old, Ruth Lawrence.

Other aspects of popularization were covered by the bookshow and the shop which sold a wide range of booklets, posters, etc, and by the finals of the annual competition for senior high-school pupils drawn from the neighbourhood of Leeds. Teams of students had been asked to work on two of three open problems, on population dynamics, coding, and tessellating, over a period of six months or so. Now, at the final, they had to present their findings using poster and micro displays, and to answer questions on their work posed by conference participants, the general public, and the three international judges whose identities were at that time not revealed.

The Pop Maths Roadshow, then, acted as a marvellous centrepiece for our work. Alas, its presence also served to emphasise the problems which face us: the inadequate financial backing which exists for such work (and the consequent additional load this places on those involved), the enormous problems of staffing and over-seeing which arise, the difficulties in obtaining appropriate media coverage,..... Yet these difficulties were overcome and about 20,000 visitors to Leeds had their conceptions of mathematics widened. More than this, the nucleus of the roadshow commenced an eighteen-month tour of the UK immediately after the Leeds meeting. Numerous locations are to be visited and, doubtless, many thousands more will have a chance to view the exhibition and to attend and participate in other mathematical events. Many thanks are due, therefore, to those who made this possible and who set high standards for others to emulate or attempt to surpass.

ICMI must, then, record its thanks to all responsible for mounting the Roadshow which accompanied our study. We are also most grateful to Geoffrey Wain, his colleagues at the University of Leeds, and to the university itself, for the help which they provided. Here a special word of thanks is due to Alan Slomson and Len Smith, who on the Monday evening organised the greatly enjoyed, mathematical nightclub, Chez Angélique. In addition to those who helped in the organisation of the conference, we should also like to thank most sincerely those bodies which made grants towards its costs, in particular, UNESCO, the International Mathematical Union, and the International Council of Scientific Unions and the Department of Mathematics at Southampton University, and, specifically, Mrs Beryl Betts, for their help in the preparation of these Proceedings.

Finally, we should like to acknowledge the enormous input of Henry Pollak to this study. He supplied the initial impetus for the work, drafted the first proposals, and his 'summing-up' talk provided the frame for our 'Study overview'. We are most grateful for his invaluable contribution.

A Study Overview

GEOFFREY HOWSON AND JEAN-PIERRE KAHANE

1. WHY THIS STUDY?

There is a general problem about the popularization of science. Science is developing very fast, yet public understanding of science is only growing slowly. Science is involved in new technologies and, as a result, in the everyday life and work of almost everyone in the world, but it appears very distant and inaccessible to most people. Science occurs or should occur in the decision-making of nations and of local bodies and institutions, and an informed citizen, whatever his or her occupation, should have some understanding of the crucial points on which these strategical choices are or should be based. However, there is now an increasing divergence between the advancement of science and the general scientific understanding of the vast majority of human beings.

Popularization of science therefore is a democratic and economic need and the provision of it may well be one of the decisive social challenges in the future.

What about mathematics? Mathematics is taught at school, it occurs at every level of higher education, it interacts more and more with other sciences, technologies and industries. Everyone knows the universal value of mathematical formulas. If 2+2=4 is such a familiar example, it is not because it is true or false, but because it is understood exactly in the same way in all parts of the world. Everyone has also some knowledge of the efficiency of mathematical symbols and figures : certainly 93+6=99 is more readily understandable than ninety three plus six equals ninety nine (to say nothing of the German or French versions).

However mathematics as a science is little regarded, certainly it is more ignored by most people than any other science. There is a specific image of mathematics, there are special difficulties in its popularization, therefore a special need for this study.

The aim of this study is to identify problems and possibilities and to provide material, not to give solutions. This is a common feature of all ICMI studies. The peculiar aspect of this study is that it is not directly concerned with educational systems. Popularization in its widest sense - any effort to bridge the gap between science and public understanding of science - certainly should concern educational institutions, and we shall return to this question. However there is an urgent need to attack the large field situated out of the educational system, partly because of the failure of the educational system itself, partly because other aims and other means have to be considered. Compared with normal teaching, popularizing is a mathematical activity where the providers are free to choose their own topic and methods. Compared with institutional learning, it can be more readily regarded as a free mathematical activity, not one of compulsion.

2. HOW DO WE KNOW THAT WE HAVE A PROBLEM?

There is no doubt about it. In most developed countries the public image of mathematics is bad. Jokes appear in the newspapers; stereotyped, incorrect views on mathematics abound. "All problems are already formulated", "Mathematics is not creative", "Mathematics is not a part of human culture", "The only purpose of mathematics is for sorting out students", "Mathematics may be important to other people, not to me"....

Even when it seems positive, the image is usually wrong: Mathematics is always correct, providing absolute truth, solid and static.

The image of mathematicians is still worse: arrogant, élitist, middle class, eccentric, male social misfits. They lack social antennae, common sense, and a sense of humour.

This is not new. "Mark all Mathematical heads which be wholly and only bent on these sciences, how solitary they be themselves, how unfit to live with others, how unapt to serve the world". This view of mathematicians, expressed by Roger Ascham, 16th century educator and tutor to Queen Elisabeth I of England, is one which is echoed in many later writings. Blaise Pascal, who was himself intimately concerned with mathematics, used to contrast "esprit de géométrie" (a mathematical mind) and "esprit de finesse" (an accurate mind). The latter was an attribute of "honnêtes gens" (nobility and high bourgeoisie), whereas the former was poorly regarded. This contrast has been a favourite theme for dissertations in French high schools, and has contributed to the view of mathematicians as strange characters, divorced from the real world.

It happens that mathematicians often reinforce this view by their behaviour or their writings: there is no shortage of examples, many of them famous. However there are figures of a very different type (think only of Sonia Kowalewska!) and, of course, most mathematicians do not stand out from the crowd.

People may have very strong feelings about mathematics and their relation to it. Mathematics yields emotional experiences. It can be seen as fun and exciting or as repulsive. It can be threatening, it can lead people to seek security in it or away from it, or it can just be dull. For some people, mathematics is a big blank, or something to be avoided, if at all possible, in the future.

As a school subject it is needed to pass examinations. It often has high status because of the intellect it seems to require; failure in mathematics may lead to a loss of self-esteem.

Many people believe that innate ability is needed in order to do mathematics. Interesting are the data gathered by the "50 lycées" project on the views held by French students. About 35% of them considered that mathematical ability is a natural gift rather than the result of training and working. Amongst those, 70% of the boys considered that they are gifted, but only 40% of the girls felt similarly.

However, these observations do not apply in the same way in every country. There is an enormous contrast between the views expressed by M. L. Sturgeon, an applied mathematician describing problems and perceptions surrounding mathematics and mathematicians in the American commercial workplace ("an obvious image problem is the absence of an association of "success" in some publicly acceptable form - acclaim, wealth, knowledge, medals, etc...- with mathematics or mathematicians... Too much worth-weight is given by our culture to superficial high profile activity. The relatively isolated and quiet activity of thinking, especially analytically, is scorned by those who cannot do it or cannot see a profit in it"), and those of D. Z. Zhang, H. K. Liu and S. Yu (who, in their paper, describe the legendary figure of Hua Loo-Keng in China, and write of public honours and consideration, "mathematics is regarded as a symbol of intelligence in China... most of the presidents of the well-known universities in China are mathematicians... some mathematicians are also political figures").

A point to emphasize is the crucial role of the image of mathematics among teenagers, precisely at the age (16-18) when they make decisions on future careers. According to Sturgeon again, for most high school students, mathematics is a _subject_, not an _occupation_. Moreover, their relations with mathematics can discourage them from going on to any kind of scientific studies. While all countries need more students in mathematics and more students in science, a bad image of mathematics may result in an enormous national loss in the near future. Conversely, a good or improved image may prove immensely beneficial to any nation in the world.

3. WHAT IS POPULARIZATION?

Let us try to find the main features that characterize this activity we call popularization of mathematics.

In the first place, it consists in *sharing mathematics with a wider public*. Each one of the words here should be understood in a broad sense. *Mathematics* can mean every subject of interest to the mathematical community, the contents, history, evolution, impact and applications, its problems, and no topic should be excluded a priori. *Wider public,* the subject of popularization, could even mean the research mathematician as the recipient of interesting and useful information outside his field of expertise. No section of the public should be excluded: children of all ages, workers, citizens, all types of professionals, other scientists. All motivations have to be considered: professional interest, curiosity, general knowledge, also prejudices and fears.

However this definition does not describe the whole dynamics of popularization. For, it also includes *encouraging people to be more active mathematically*. Mathematics is not so interesting as a collection of results as it is when considered as a way of thinking: how to formulate a problem, how to look for a solution, how to demonstrate. In a way mathematicians have to be more ambitious than other scientists, because the main value of mathematical concepts is to be used, not only grasped or understood. There are good examples: certainly G. Polya succeeded in popularizing widely his approach to mathematical investigation and discovery. Christopher Zeeman insists on the choice of theorems and proofs for a popular talk: theorems should be *noble* (capture the quintessence of some mainstream branch of mathematics) and the proof should be *elegant, rigorous,* and expressed in a *few lines.*

Now comes a question. Is popularization not the purpose of mathematical teaching? The specific feature of popularization is to provide a *mathematical activity in freedom, not by compulsion,* as we already said. It should not imply work and hard effort, but freedom and pleasure. How far that is from school activity will be discussed later.

From a more general point of view, popularization of mathematics consists in bringing (or bringing back) *mathematics into human culture or, rather, cultures.* For, it is vital to recognise that we are not just writing here of the shared culture of middle-class, white Western Europeans and North Americans. This point, though, is made more strongly in the report of the Working Group on 'Mathematics in different cultures' and in the paper by Gordon Knight which are reprinted in this book. The cultures of our time need a scientific understanding of the environment and technologies, with constant changes of scales, programs, controls, numerical data, statistics, forecasts, models etc, and nobody should be afraid of the few mathematical notions involved. On the contrary, mathematics together with its history and present applications forms a natural link between humanities and technologies.

Maybe it is possible to go on and find other features of popularization. But it may be appropriate at this point to observe that we have already reached a contradiction. To bring mathematics into human culture means mathematics for all. Free activity means mathematics for those who like it. More activity means mathematics for those who are able to be active. Sharing with a public means mathematics for those who are willing to hear.

In order to deal with this contradiction we have to consider the people and matters involved in more detail: to whom, what, by whom? Let us begin with a special but crucial aspect of the question.

4. MATHEMATICIANS AND MATHEMATICS

There were three main talks on the first day of the symposium. Christopher Zeeman started from his experience of the Royal Institution lectures, work done with the BBC, and a series of mathematical Saturday mornings with gifted students. He advocated giving some proofs, even on TV programs, and for choosing beautiful, "noble" theorems (eg $\sqrt{2}$ is irrational, Σ $1/n$ is divergent). He developed the interplay between continuous and discrete phenomena, continuous and discrete theories, with these self explanatory examples:

Mathematical Model

		D	C
Physcial phenomenon	D	Dice Probabilities	Planets ODEs
	C	Music Harmonics	Waves PDEs

He gave us an idea of the topics which he had used on the Saturday mornings: spherical geometry, perspective, gyroscopes, gears, knots, curves with constant diameter. And he had time enough not only to give a few proofs on spherical triangles and gyroscopes, but also to demonstrate 'precession' through the use of a home-made boomerang. The talk was a fine example of popularization in action, with personal views on classical matters.

Lynn Steen considered recent developments in the USA in the relations between mathematicians and journalists. (See the paper reprinted later in this volume.) He elaborated on the necessary distinction between the new and the news. Mathematicians are interested in new results (say: Falting's, de Branges'), new trends (eg: dynamical systems, fractals), new points of view (eg: numerical, algorithmic). Journalists are interested in news, "what the public pays for". He described how it was possible for the new to be converted and sometimes changed into news. There is now a Committee of US mathematicians, representing 15 associations, in charge of the relations between the mathematical community and the journalists, which was producing better and a deeper understanding from both sides. It was a talk which was listened to with much attention.

Alain Connes had been asked to give personal views on some recent trends in mathematics. He chose to speak on the relation between pure mathematics and quantum physics, and the limitations imposed by classical set theory. Actually he had in mind his own recent work and those purely geometric situations where classical set theory is useless. The most interesting examples come from the interplay between geometry, algebra, and a phenomenon in physics (the quantum Hall effect) "begging for a mathematical explanation". His talk, therefore, ranged widely, first moving from curvature of surfaces to experimental results where bizarre plateaux appear. He then showed that the plateaux express the same kind of discretization as the integral of curvature in geometry, and share a common explanation in non-commutative geometry. Here quantum sets entered and the necessity to revisit the axiom of choice. This was an ambitious talk and was received in two ways. Part of the audience, mathematicians and non-mathematicians, was fascinated and very happy, many reacted negatively, particularly at the end of the talk. It was a very different reception from that which Connes encountered when giving a slightly less ambitious talk in front of a very large audience in Palaiseau (France) - the meeting which Emmer describes in his paper. Was it that part of the audience in Leeds was not so much interested in grasping new ideas in mathematics as in discussing ways to popularize that mathematics with which they themselves were familiar?

To some degree, the answer to this last question must be 'Yes'. Yet the lecture also highlighted other vital issues concerning popularization. The notion of a 'contrat didactique' between pupils and teachers has been much discussed and refined in recent years. As this and other lectures in the Pop Maths Roadshow demonstrated, it is a notion which also holds good for lectures on popularization. The audience has expectations and limitations and it is essential that these are appreciated by the speaker (see, eg, the paper below by Tony Shannon). Alain Connes' lecture was especially valuable then, not only for the way in which it impressed and was appreciated, but also because of the way in which, particularly in retrospect, it raised many issues - alas, not discussed and developed fully - about the problems of popularization at all levels.

Such considerations lead us to two fundamental topics: 1) the need for, and ways of, popularizing mathematics among mathematicians and mathematics teachers; 2) the role of research mathematicians in the process of popularization. Let us summarize a few aspects of these questions.

a) The image problem. In most countries, it appears that the popularity of mathematics diminishes during the time that students are exposed to mathematics at school. Other school subjects do not seem to have such a severe image problem. Mathematics may seem more esoteric and less obviously useful than other subjects - even to mathematicians themselves, who may be in need of popular explanations of the work being done at the frontiers of their subject. There may be something of a paradox in attempting to popularize a subject whose own practitioners are often ignorant of current developments.

b) A positive evolution. The relation between research mathematicians and mathematics teachers is improving in many countries. There are good and stimulating papers by research mathematicians on current developments in mathematics in many professional journals. Expository papers are more and more appreciated and some journals have a very good policy of such invited papers. In Congresses of Mathematicians the style of the general talks has changed for the better in the past twenty years: to be understood by the audience appears now to be a more widely accepted aim.

c) An increasing but still insufficient interest among mathematicians. Popularization may be needed even from the point of view of research. Let us quote J. Ekeland:

> Pour faire de la vulgarisation [popularization], il faut être capable de dire ce qu'on fait et pourquoi on le fait. Cela contraste avec une certaine pratique des mathématiques - combien de

chercheurs ne s'élèvent pas à une vision globale, et se contentent de résoudre les problèmes qui passent à leur portée?

Une erreur commune parmi les mathématiciens est de croire qu'un résultat difficile est nécessairement important. Pour faire de la vulgarisation, il faut être capable de discerner ce qui est fondamental - c'est bien souvent décapant par rapport à la pratique courante de la recherche.

d) The historical point of view. Mathematicians have a much stronger relation with the past of their science than other scientists. Great works of the past retain their value in mathematics far more than in, say, chemistry and physics, and at the same time they may have a human value for other people. History of mathematics provides a way to popularize mathematics of the past. Moreover, popularization of recent works or trends in mathematics can be considered as a piece of recent history.

e) The philosophical point of view. Mathematicians can be very interesting to other people when they ask themselves the meaning of what they do. Are they exploring and discovering new continents, are they inventing and building new machines? Are mathematical objects given in nature or are they a construction of the human brain? Are they changing in time, and how? Are definitions a starting point or a historical achievement? How do mathematicians think, with images, words, or anything else? How are they inspired by the external world, by social needs, by applications? What do they call models, and what, theories? Is there any way to explain what the physicist, Wigner, called the unreasonable efficiency of mathematics in the natural sciences?

It is hoped that these fundamental questions will be discussed among mathematicians, in particular on the occasion of their International Congresses. Popularization would be unfortunately biased if active mathematicians took no part in it.

5. TO WHOM, WHAT, BY WHOM

There are very strong and opposite opinions about the range of popularization. To what proportion of people can mathematics be popularized? What is the proportion of mathematics which can be popularized to a large audience? Answers to both questions vary from 10% to 100% . Actually these proportions have no objective basis and express only personal feelings. In what follows let us assume that we do not exclude any part of the populace nor any part of mathematics from the processes of popularization.

A good way to begin, suggested by Henry Pollak, is to build a grid, with 4 columns: the kind of people, their motivation, the kind of topics, and the characteristics to take into account. In the first column you may have two big groups: adults and children. Among adults you may define a number of subgroups: colleagues, informed citizens, workers, parents, the retired. Among children, older ones who like mathematics, and, regrettably, older ones who don't; and younger ones who are still indifferent or undecided. On each line you may try to fill the other columns.

Since we have already discussed the possible content of the table entry headed 'colleagues', let us look at that for informed citizens. What is their motivation? They read news, and they want to participate in important issues. Which topics can be appealing to them? What knowledge and understanding will help them to take, challenge and appreciate decisions more effectively? How can they be helped better to comprehend the contribution that mathematics makes to social well-being? How can they be reached? As we shall see, the range of media is wide and contacts can be made in a variety of environments, even the shopping centre.

Another interesting and rapidly expanding subgroup is that formed by the retired. They want to enrich their lives, and also to be able to talk with grandchildren. They can be interested in modern trends, history, and culture.

Their characteristics are to have time, to like travelling, to be able to go to lectures and exhibits.

Let us add a new entry: elementary school teachers. Very likely their motivations and characteristics are different in different countries - this is also true of informed citizens and the retired. A common feature is that they have to give answers to children. They may like to have good and simple examples about new terms or new data. Unfortunately, a characteristic is that often they are much more knowledgeable and experienced in other subjects than in mathematics, and this may lead them even to overlook the claims of mathematics as part of human culture.

Let us look at the third column: the kind of topics suitable for popularization of mathematics. Since no topic should be excluded, a list would be irrelevant. The selection depends on the recipient according to a few simple principles:

- The *relevance* of the topic to the actual problems of society or of groups within the society. Examples : the need to appreciate statistical information (eg sample polls during an election campaign), to evaluate the real magnitude of the interest to be paid on a loan, etc. (See Nemetz's account of the recent popular demand in Hungary for information on percentages.)

- The extent to which the topic introduces and illustrates *methods* which have value for general (systematic) problem solving, such as algorithms, decision trees (eg for puzzles), heuristics, and approximations.

- The potentiality of the topic to cause *positive emotional attitudes* towards mathematics (its beauty, its universality, its unexpectedness, its reliability, the effectiveness of a formula,...). To a certain extent, the considerations raised here rest on "primary instincts", somewhat analogous to those which generate an almost universal interest in cosmological questions in astronomy.

- The potentiality of the topic to help *others understand* those people who have to deal with mathematics "professionally" (at various levels!).

Examples: policy makers must be able to understand the analyses of statisticians, the claims of curriculum makers, the productivity reports of research mathematicians; parents have to understand the problems of their children in their mathematical education.

An important idea is the relation between mathematics and arts. The presence of mathematics in various artistic domains, such as painting or sculpture, has been treated in numerous places. It is interesting to note that this question still raises interest, as is testified, for example, both by the exhibition of John Robinson at Leeds (part of the Pop Maths Roadshow), as well as by the project actually under completion at the Musée des Beaux-Arts de Montréal of preparing a series of activities for primary school children (age 11-12) based on various pieces belonging to the museum. The aims underlying such an initiative from a museum can be many: to make the museum collection better known, to alert those people primarily interested in the arts to the role that mathematics can play in helping to structure space or, by providing interesting shapes to work with, to provide an unusual field of application to mathematically-inclined people, etc.

As a specific example, the kaleïdoscope is a particularly successful example of popularization, but only to the extent that the attraction caused by the intrinsic beauty of the kaleïdoscopic image is used to induce people into the building of an appropriate geometrical model.

Various other topics could be mentioned here, which could be presented to most of the possible audiences described above in appropriate ways. To name a few: knots; cryptography; magic squares: non-linear dynamics (catastrophes, fractals); logic; probability; statistics; optimization;... In fact, as is demonstrated by recent, apparently successful books popularizing mathematics (eg K. Devlin, *Mathematics*: *The New Golden Age*; I. Peterson, *The Mathematical*

Tourist; I. Stewart, *The Problems of Mathematics*), almost any mathematical topic can be presented at a level appropriate to the target audience.

An important point resulting from the above considerations is the great diversity of mathematics. If an individual is not interested in a certain mathematical topic, this does not imply that he should avoid any mathematics. There are many other topics to choose from, and some of these might have considerable appeal to that person. (Think of music: no one is expected to appreciate everything that is described as 'music'. Moreover, it is generally accepted that certain types of music are somewhat esoteric and call for considerable "experience" in order to be fully appreciated.)

For those who take part in popularization, there seem to be some common features.

Their aim is not to provide complete information on any subject. Rather, their role is to invite people to go further. As an example, when organizing an exhibition, you will pay considerable attention to the selection of books which can be purchased by visitors on their way out.

They cannot tell the whole truth, but what they tell should be part of the truth. People should be able to rely on what they hear, read or see, and not in the future be obliged to unlearn before proceeding further.

They have to face the general law, that the level of a talk is likely to vary inversely with the size of the potential audience.

However there are different views and different styles. For example, should the recipient be mentally active? For some colleagues the answer is obvious and positive, they propel the audience into a position of active participation and try to share mathematical thinking (see, eg, the paper by Hatch and Shiu). For others the answer is not so simple. They consider it their duty to provide help to those who want to learn *about* mathematics, either because they are too weak or too lazy to try to do anything in mathematics. Henry Pollak argued in favour of a different form of laziness, in that one purpose of mathematics is to find simple ways of doing things that were previously more difficult.

Popularization should involve mathematicians, including very good research mathematicians, as we already said. It should involve mathematicians in a wider sense: professionals, educators, teachers, even students. However it would not go very far without the help and active participation of all kinds of professionals within journalism, radio, TV, museums, etc. In particular, scientific journalists have a crucial role, and the quality of their information is one of the obvious challenges of popularization.

Let us express another idea. Popularization should involve scientists who are not mathematicians. In the study on "Mathematics as a service subject" we quoted biologists who were interested in having their students acquainted with mathematical thinking. If they were able to explain why they think this would be valuable, in what circumstances they are faced with problems where a mathematical approach would prove useful, then this itself would be an excellent piece of mathematical popularization.

We can use beautiful writings by physicists, from Galileo Galilei to Richard Feynman, where mathematics appears in its full power, glory, and simplicity. Similarly, with philosophers, from the great Plato to Gaston Bachelard or Michel Serres (who provided the French public with new editions of classics like Père Mersenne).

In the popularization of mathematics, there is a place for many actions and many actors.

6. POPULARIZATION AND SCHOOLS

We have already emphasized some important differences between popularizing and teaching. Popularization has a wider range (audience and topics), many more ways and more freedom. Lynn Steen points out an essential difference in the aims: *the purpose of popularization is to raise awareness, not to educate, and the criterion of success is not an increase of knowledge, but a change in attitudes.*

On the other hand, we mentioned as a fundamental problem the need for, and ways of popularizing mathematics among teachers, and we considered school children and elementary school teachers as examples of particular targets for popularization. Let us be more specific now about students of all ages.

1) Older students who enjoy mathematics might profit by learning more about (and not necessarily only more *about how to do it*). The information given to them should indicate how mathematics is evolving both in itself and in the range of its applications. Part of the information for these students should concern their future careers, stressing job opportunities in mathematics and in domains strongly related to mathematics.

2) There is a need to popularize mathematics among weak (older) students, if only to convince them of the importance of acquiring basic skills for their everyday life as well as for jobs in which mathematics can play some rôle, even secondary. The main purpose is then to have these students gain self-confidence in their mathematical capacities and to help them develop a positive attitude to it. Attitudes of fear and of inadequacy will have built up over many years of conventional schooling. Almost certainly something unconventional, something taking place outside of the classroom, will be needed if these attitudes are to be changed. Only then are they likely to come to perceive the mathematics studied in school as understandable and non-threatening.

3) It is probably fair to say that no child entering elementary school dislikes mathematics. Popularization at this level thus aims at preventing such a dislike being generated through school mathematics.

Now we have to question our starting point. If we consider popularization in its widest sense - any effort to bridge the gap between science and its public understanding - it should include education. Until now, we have considered popularization as complementary to, indeed in some senses a correction for, the educational system, and we have succeeded in expressing specific aims and features. However, looking back at what we suggested for school children, is it not true that part of it should be done at school?

School mathematics need not necessarily be boring, nor schools out of touch with real life. Exhibits and projects can be realized in the schools. To consider one or two examples, the competition for students from the Leeds area which was 'judged' at the Pop Maths Roadshow is a very successful way of *encouraging young people to be more active mathematically* (they work in teams, for months, on open and difficult mathematical themes), and some French "projets d'action éducative" have had a very good impact in *bringing mathematics into human culture*. The mathematical trails organized in several Australian locations and described in Leeds by Dudley Blane could very well form part of regular education. Collections of short pieces of TV recreational programs could be used in class rooms.

Generally speaking, if a means of popularization is successful, one should consider translating it into regular education.

This idea, if applied, may bring new life into, the teaching of mathematics. First, (and this would be true also for other subjects) it is good to introduce open activities and non classical themes. Secondly, and this is most important for the teaching of mathematics, it would contribute to making mathematics appear as a living science and not merely a collection of techniques or a universal language. The Japanese approach to *mathematical literacy* is to build a whole curriculum where the students have to learn something *about* mathematics without learning mathematics in the usual sense; it is part of the spirit of popularization in the curriculum itself. The need for historical comments, connections with other sciences or technologies, experimentation, modelling, goes in the same direction.

Therefore, we have to think of popularization before, during and after regular education, and try to reduce the opposition between both activities. As Victor Firsov expressed it, a common feature of both good popularization and good education is to promote self-esteem.

If mathematics teachers are involved in popularizing mathematics, at school or outside, it may be expected that their educational efficiency will grow. They need information and material and many of their associations are aware of the problem. Professional training should include consideration of this aspect, and mathematical educators should bear in mind the question of how to connect future teachers with mathematics as a living science. In bridging the gap between mathematics and its public understanding mathematics, teachers have an essential role, perhaps outside, but certainly inside the school.

7. THE MEDIA AND THE MESSAGE

We have already mentioned various means by which mathematics can be popularized. Traditionally, the two main media were the written and the spoken word. Here, it is important to realise just how long a history popularization has. 'There are some, King Gelon, who think that the number of the sand is infinite...' begins Archimedes' *Sand-Reckoner,* written over two thousand years ago, and intended to reveal more clearly concepts of number and comparative magnitude. Here we have an early example of an author attempting to demonstrate to his ruler and others in 'high places' the power of mathematics, and, doubtless, the need to make adequate provisions for the professional mathematician. That the popularization of mathematics can take other forms and serve varied purposes is well illustrated by two eighteenth-century attempts to 'popularize' Newton's works: Maclaurin's *Account* and Voltaire's *Elémens*. Both attempt to bring Newton's work to the attention of a wider audience - but with very different motivations.

To Maclaurin, the aim of popularizing Newton's theories was not to 'excite mere wondering', nor indeed to strengthen the position of mathematics within society. Rather, he believed such knowledge to be 'the firmest bulwark against atheism'; not, it might be said, Voltaire's primary purpose as an author!

It is of course, more difficult to trace the history of the spoken word as a medium for popularization. However, the Leeds study provided us with one historical link. In the 1590s, Thomas Gresham endowed a series of public lectures in London intended to acquaint the citizens with new developments in fields which included geometry and astronomy. The first professor of geometry (who had to deliver his lectures in Latin, and then, later the same day, in English) was Briggs (of logarithm fame). These popular lectures have continued to this day, the current 'Professor of Geometry' being Christopher Zeeman. Gresham's aims might be described as being to generate awareness, and, so far as mathematics was concerned, an appreciation of how it might bring added power to London's merchants, navigators and others.

These few examples illustrate not only a long historical tradition of popularization, but also help us see that popularization can serve many purposes. Indeed, if any particular initiative in popularization is to be successful, then it would seem essential to identify with considerable precision exactly what purposes are sought in that instance. Here the choice of, and the availability of different, media can be influential.

For now, in addition to the written and spoken word, we have, for example, the possibilities offered by striking and sophisticated visual, non-verbal media, and a range of activities available which go far beyond the traditional and still influential competition (to be found in eighteenth century UK newspapers and annuals) or problem-solving (here one recalls that, for example, the wolf, goat and cabbage problem was known at the court of Charlemagne and had independent origins in Africa and elsewhere).

Working groups met at Leeds to consider the different media employed to popularize mathematics. Below we summarise some of the points made in these groups or in general discussion. More detailed consideration of certain aspects can then be found in the selection of papers which follows.

8. BOOKS AND MAGAZINES

Despite the increased competition from other media, books and magazines still play a vital role in the popularization of mathematics. We have already mentioned the work of such authors as Devlin, Peterson and Stewart, but these are but a few of the many represented at Leeds, and even these formed only a small percentage of the list of authors in print. Visitors to Leeds were able to see mathematical magazines from many countries: yet, for example, since the Leeds meeting the first number of *Quantum*, a U.S. version of the famous USSR periodical *Kvant*, has appeared and a new UK magazine, *Mathematics Review*, aimed at the 16-20 year-old has been announced.

The group which met at Leeds considered two main problems: what characterises a 'good' book or magazine, and what are the major difficulties in writing popular books or articles and in running a magazine.

Four components of 'mathematical culture' (or of a 'mathematically cultured' person) were identified:

(a) a knowledge of elementary facts and methods,
(b) the development of a certain way of thinking and of approaching problems,
(c) some knowledge of the history of mathematical concepts and theories,
(d) some knowledge of recent developments.

Regrettably, school usually only attempts to secure (a): other objectives have to be developed elsewhere and books and magazines can be powerful aids.

In particular, goal (b) can be sought from an early age. Although, it was felt that 'good' books might well focus on only one of the aims (a)-(d), a magazine (if not addressed specifically to very young children) should aim at the development of (b), (c) and (d) at least. Serious doubts were raised about magazines consisting of problems or puzzles only, which might not necessarily introduce the reader to elegant, powerful, *mathematical* approaches to problem solving. (Tony Gardiner's *Mathematical Puzzling*, 1987, Oxford, was seen as a book which successfully avoided these dangers.)

There is an obvious need for magazines and books for young children (4- 6) to provide opportunities to *play* with *mathematics*. (This link between play and mathematical activity was one which surfaced many times in discussion.) Moreover, such play should involve not only children but their parents or others. The use of mathematical play with children to affect the attitudes of adults towards mathematics was a notion underlying the US *Family Math* publications described, and on show, at Leeds. Manipulatives, cut-outs, etc. immediately raise the attractiveness of such books. Marion Walter's *Annette: a mirror book* and a whole range of books from Tarquin Publications exemplified this well, and also demonstrated how much mathematical activity and learning could be generated with only minimal text. (In this section, we refer mainly to books written and available in English, although there is no shortage of recommendable texts in other languages.)

Geometrical situations and number patterns and problems would seem to play a particularly significant role in material for older (7-12) primary school children.

It is often the introduction of algebra, the manipulation of symbolic expressions and the translation of word problems, which appears to highlight marked differences in attainment amongst junior high school students. Now there is a need to make algebra more meaningful and attractive to the average and below-average child whilst still offering something new and outside the school curriculum to those whose mathematical interests and talents are beginning to blossom. *Thinking mathematically*, by J. Mason, L. Burton and K. Stacey was thought a good exemplar of the desired approach. Another book with a good (b) (i.e. mathematical thinking) component is *A way with maths* by N. Langdon and C. Snape.

In general, reaching secondary school students is difficult because now one is faced with preconceptions and with hostility towards mathematics. Indeed, it was doubted if books and magazines can by themselves do anything to overcome such hostility. The school must play the major role if that battle is to be won. At this level magazines take a variety of forms, from those linked tightly to school curricula (often published primarily for commercial reasons) to those (e.g. *Tangente* in France) which go beyond and outside it.

Once, however, one begins to deal with pupils post-16, with some commitment to mathematics, then many new opportunities arise. Perhaps the principal aim now is to show mathematics as a live domain of research. However, particularly for those who will become teachers, there is an important place for the history of mathematics. These are aims emphasised by, for example, *Kvant* (USSR), *Delta* (Poland) and *L'Ouvert* (France), and in such books as Ian Stewart's *The problems of mathematics*.

These who do not study mathematics may perhaps at this stage start reading books on mathematics, not making specifically mathematical paper and pencil demands, but matching mathematics with philosophy, culture and art. Such books can also be read with pleasure by mathematicians. Examples are *Images of Infinity* by Ray Hemmings and Dick Tahta, D.R. Hofstadter's *Gödel, Escher, Bach: an eternal Golden Braid*, *The mathematical experience* and *Descartes' dream* by P.J. Davis and R. Hirsh, and R. Rucker's *Infinity and the mind* and *The fourth dimension*.

There is also a clear need for books for adults who wish to understand mathematics, perhaps to help their children, perhaps to come to terms with ideas which passed them by at school: H.J. Jacob's *Mathematics, a human endeavor* and L. Buxton's *Mathematics for everyman* are examplars.

Who, though, is to write popular books and popular magazine articles? It was felt that many professional mathematicians fear a loss of reputation as a result of writing popular books. Of those who do, many hesitate to write a second as a result of ill-judged criticism of the first. In the USSR there is a tradition of leading mathematicians providing popular books and papers. One notes also that the first edition of *Quantum* contains a paper by W. Thurston. It is important that links between leading mathematicians and the wider public is strengthened, and the book or magazine is a powerful medium for accomplishing this.

9. NEWSPAPERS

There can be few people with experience of being interviewed by a newspaper who have not later dreaded reading the views and words ascribed to them. Yet the newspaper remains a powerful medium which we cannot ignore. A few, 'quality' newspapers are prepared to provide a regular space for mathematics written by a mathematician. Michele Emmer describes later in this book how he sets about to compose such a column. In general, though newspapers are concerned with 'news', and this presents particular problems to a discipline in which major breakthroughs appear infrequently and then present special problems of communication because of the complexity of both the mathematics involved and also, frequently, of the mathematical motivation. Often the part which mathematics has played in, for example, scientific and technological advances is also left unmentioned because of the difficulty of presenting comprehensible accounts.

Yet it is essential that this contribution of mathematics to social well-being is widely publicised. Some suggestions on how this might be done more effectively are to be found in Lynn Steen's paper.

But what of the view of mathematicians conveyed by newspapers? Alas, it is only 'news' that a mathematician can be well-balanced and have a host of extra-mathematical interests when his or her obituary is published. It is the eccentric mathematician who commands most attention when alive. A brilliant piece of journalism, which was reprinted in many countries, recently described a leading present-day mathematician:

> "he has no home, no family, no money, no distraction from his obsessive quest for elegant solutions", "a mathematician is a machine for turning coffee into theorems",... .

In one sense it was an accurate 'picture', but taken from a particular viewpoint. It would, however, have hardly increased the number of adolescents thinking of becoming professional mathematicians! Yet how is the image of mathematicians as boring and dull or madly eccentric (one thinks again of, say, the most impressive book and play about Alan Turing and the beautiful TV film on Ramanujan) to be overcome? The image of the mathematician as a fashionably dressed female (see Celia Hoyles' paper on her TV series) is an important step in breaking down traditional, stereotyped expectations.

10. TV AND FILMS

It is hard to over-estimate the influence of TV as a medium, and the image of mathematics projected by TV is crucial in any discussion on popularization. There are, of course, many 'educational'/'instructional' TV programmes but these are usually (although not always) directed at a specific and often limited target population. In addition to these, though, are many other programmes which influence the viewers' conceptions of mathematics.

Four particular categories of audience for whom 'popular' television is already providing programmes were identified.:

(A) Children of primary school age (5-12) when at home

Example: Square One (Children's Television Workshop, New York - see the paper by Joel Schneider and Edward Esty). The rationale is that watching TV itself is fun and here there are 'fun' characters evidently enjoying mathematics, *ergo* doing and watching mathematics can be enjoyable.

(B) General (family) audience

Example: Fun and Games (Yorkshire TV, UK - see the paper by Celia Hoyles). Solving puzzles is fun. If nothing depends on a participant's success or failure (no prizes, points, or enhanced status), then no one is discouraged from 'having a go'. Enjoy pitting your wits against the non-expert TV participants. Basic quiz show popularity remains but the emphasis is now more on the puzzle than on 'success' (in the form of prizes). Humour is provided by the presenter in order to retain interest and there is always the strong possibility that viewers will afterwards ponder on some of the puzzles/solutions.

(C) 'Lay' (non-professional) audience interested in applications of science, and technology, and possibly interested in scientific concepts for their own sake.

Example: Horizon (BBC,UK), Equinox (Channel 4, UK). Scientific applications are important - for socio-economic reasons, or as pure advances in scientific understanding. It is also the case that many mathematical topics, e.g. chaos, can lead to extremely beautiful and tele-genic pictures.

(D) Parents interested in their children's education

Example: Help your Child with Maths (BBC, UK). Many parents wish to understand more about what is going on at their children's schools either in order to assist the children or simply to ensure they still retain links with them.

It was noted that producers working on Category C programmes which were largely or wholly about mathematics tended to concentrate on 'pure' mathematics, but that when 'applied' mathematics underpinned the content or procedures of a 'science' programme (e.g. the Equinox 'chaos' programme) then mathematical details (numbers, formulas, equations) were rarely shown. The need to give greater emphasis to the work of applied mathematicians and to the mathematical ideas they employed was expressed.

Classification by audience is, however, but one possibility. Another classification would be by the type of response envisaged. Is the viewer being asked to participate in 'problem-solving', or is it the aim to improve his or her appreciation of mathematics by providing information on its history, culture or new trends in an interesting form?

Again, is the strategy of programme design to be the implantation of some mathematics into the traditional clichés of TV culture, or can the issue be attacked more directly by finding a way to present the subject enjoyably and then hope to have this accepted by sponsors, TV authorities, etc?

There is no doubt that selling the idea of mathematics programmes is not easy. Neither, for that matter, is the essential co-operation between professional mathematicians and the 'communicators', in particular, professional film-makers, as finely attuned as one would wish.

These are just two aspects which could gain substantially from international exchanges of information and experience. There are now many mathematical shows on TV around the world. Given the importance of the medium and the influence it can have on children and adults, it would seem crucial to devote more consideration to how mathematics can be projected more accurately and effectively. Certainly, there would seem scope for an ICMI study specifically devoted to mathematics on television.

11. EXHIBITIONS

As we have already written, the Leeds seminar was complemented by what was possibly the largest mathematical exhibition ever to be mounted. We were also fortunate to have at the seminar, Roger Lesgards of the Paris 'City of Science and Industry - La Villette', and Michel Darche who has been much involved in that museum's travelling exhibition, Horizons Mathématiques.

Participants heard, with some envy, of the financial support which La Villette receives from the French State, and also of the particular problems of establishing a 'permanent' exhibition which does not quickly ossify. Large mathematical exhibitions are comparatively recent entrants on the scene. Many national museums still do not have collections featuring mathematics. It is possible that one of the main reasons for this, and also for the fact that so few mathematicians have places on the boards of museums, is that the mathematics community has failed to overcome the credibility gap that the subject can be portrayed as attractive and entertaining.

That it can was clearly demonstrated at Leeds. (A bigger question, still to be answered, is how one bridges the gap between what children and adults find attractive and entertaining mathematics and mathematical activities in exhibitions and, on the other hand, what is to be found in classrooms, lecture theatres and textbooks. One is not seeking here the facile solution that the latter should more closely resemble the former. Certainly, institutional mathematics has lessons to learn in presentation, but, say, the acquisition of essential techniques which must be emphasised in institutional mathematics learning is unlikely ever to be given such weight in an exhibition.)

There is an essential need, then, for mathematics societies to liaise with museum directors with the crucial aim of ensuring that mathematics becomes better represented in national and other large museums.

But there is also a strong case to be made for local and for touring exhibitions. Horizons Mathématiques has already travelled to many countries of Europe, Africa, Asia and the Americas. Such exhibitions can provide a centre-point around which many other kinds of mathematical activities can take place or can develop. Nor need these always be mounted in traditional 'educational' institutions or contexts. The Pop Maths Roadshow is to visit two of England's cathedrals; the exhibition 'Science et Contes', mounted by Laval University, Quebec, has visited shopping centres in various small cities in the province. The latter includes mathematical examples from bifurcation theory, fluid dynamics and fractals. Moreover, because it visits small cities where it has fewer competitors it can become a major attraction. The genesis and aims of another touring exhibition, on the theme of knots, is described below in a paper by Brown and Porter.

As with the other media, we are faced within exhibitions with the need both to provide information and also to supply activities through which mathematical understanding can be gained. Neither aspect can be safely neglected. There *is* a need to inform about what mathematics and mathematicians can do, but there is a need to involve visitors in some types of mathematical activity (without conveying the impression that such activities are in any way canonical).

As in the case of TV, there was much discussion about the respective roles of the professional mathematician and the professional 'exhibition designer' when it came to mounting exhibits. The need for the former to give the lead was emphasised, as also was the need for him or her still to maintain some influence as the balance of direction later shifted towards the professional 'exhibitor'.

Activity, colour, beauty, etc. should be there, but should never be allowed to supplant serious mathematical content and purpose. Ideally, too, the exhibition will never be a 'stand-alone' medium. If school pupils are to visit one, then there should be preparatory tours or material available for

A Study Overview

teachers and, whenever possible, suggestions for follow-up activities. We have also already stressed the importance of having books, construction kits, etc. available for sale, so that interest, once aroused, can be further developed.

12. GAMES AND PUZZLES

Games and puzzles play an important role within mathematics and in its popularization; a fact well documented in Miguel de Gúzman's paper. Internationally published collections of mathematical puzzles or 'recreations' intended for a wide audience have a three-century-long pedigree from Ozanam to Martin Gardner. Individual puzzles, many connected with exponentiation (e.g. rice grains on a chess board) can be traced back many, many centuries. Clearly they do excite considerable interest and have the capacity to encourage and inculcate mathematical thought. Attempts were made at the seminar to distinguish between 'games' and 'puzzles'. The working group distinguished between them largely on the grounds of 'competition' and the number of people involved: thus 'puzzles' became 'one-mind' games. Another distinction was provided by Tony Gardiner : puzzles (e.g. tic-tac-toe/noughts and crosses) have a mathematical solution which, once mastered, make their playing of little interest. 'Games' either present far too many possibilities to be mastered mathematically (e.g. chess), or introduce a random element (e.g. bridge, backgammon) which adds spice.

It is, of course, very hard to tell exactly what influence playing certain kinds of game will have on a person's future development. In many cases that influence can be readily seen - and there is no doubt that they serve to foster mathematical thought. In other instances it would seem that the bridge has never been made which allowed the games player to transfer to 'mathematics' his or her ability to appreciate logical problems and structure and to develop strategies. Game playing develops certain attributes needed by 'mathematicians', but some connection must be made if those attributes are to lead to mathematical success and/or an interest in mathematics. Thus, to take a particular example, computer games can develop spatial sense and awareness. However, these attributes (like, e.g. the ability to appreciate logical patterns and structures) are not those solely of mathematicians - for architects, say, must possess them too. Geometry is by no means synonymous with 'spatial awareness'. Unless, then, specific attempts are made to mathematise situations or to provide links, i.e. to identify and capitalise upon possibilities, then much of the potential mathematical value of games playing will be lost. But, will too much mathematical introspection lead to a loss of enjoyment in games-playing? Finding a balance is not easy - care and patience are needed.

Although, many instances can be given of games providing a context for the consolidation of mathematical content, perhaps they are most valuable in supporting the development of the processes of doing mathematics. For example, chess provides experience of an axiomatic system - there are objects which can be operated upon in certain ways. Again, in some computer games the exact functions of certain keys, although well-defined, are not disclosed *a priori*. They can only be discovered through trialling, conjecting and testing.

The use of games and puzzles is, therefore, not without its problems. On the one hand, the value of games might be underestimated by those who take the view that playing them 'is not real mathematics'. On the other, it is easy to allow the playing of games to become an end in itself - or to believe that any mathematical learning will occur naturally by osmosis.

13. COMPETITIONS

Mathematical olympiads have a long history in many countries and the International Mathematical Olympiad now attracts from over fifty countries. The IMO certainly has the effect of drawing attention to mathematics and in this sense, as in some others, it has a role to play in the popularization of mathematics.

However, the IMO and, in many cases, National Olympiads are 'preaching to the converted' so far as actual participants are concerned - their effects can in certain circumstances even be negative. As a result, much attention has been focussed in recent years on competitions for 'the masses'. Some, such as the Australian Mathematics Competition, have enjoyed phenomenal success: after less than ten years in existence, the ANC attracts over 400,000 entries - over 80% of all High Schools join in, and more than 1 in 4 of all Australian High School pupils take part each year.

Attention at Leeds, then, tended to be concentrated on non-Olympiad competitions and on exchanging information on the different ways in which these can operate. Papers by Burjan and Vrba, and Nemetz, reprinted in this book, describe some such activities in Czechoslovakia and Hungary respectively. Other competitions described at Leeds included some based upon Liverpool and Birmingham in England, Kuwait and South Africa. These targetted different age groups, 11-12, 13-14 and 15-17, and used a variety of methods and media (e.g. radio in the case of Liverpool). Problems were often of the 'take-home' variety and great stress was placed on the prize-giving ceremonies at which students were invited to describe their work and, for example, take part in team events, such as relays, with parents present (and, on occasion, joining in the mathematical activities).

Subsequent discussions covered the nature of competition problems, the categories of people involved and the dangers created by competitions.

It was thought that the problems should not be too curriculum-driven, but reflect an authentic mathematical experience. While a multiple-choice format may be efficient for large groups, there should be questions requiring extensive work and good exposition. We should assign tasks or projects, as in the case of the Leeds competition described earlier. In any case, care should be taken to see that questions are appropriate to the group and embody variety and imagination. A take-home contest relieves pressure and allows for reflection and exploration without, in the experience of those who have tried it, compromising the integrity of the competition.

The groups to be considered in such activities are the children themselves, their parents and teachers, academic mathematicians, sponsors and the public. Eleven to fourteen seems to be a good age for children. They have a natural enthusiasm and their interests have not yet narrowed; but they are old enough to cope with the competitive side. There should be opportunities for social interaction, perhaps through participating at a common place, and certainly at a prize-giving ceremony, where there should be many awards. Children might be encouraged to give short talks. At school, the contest can be the instigator of extracurricular mathematical activity.

Parents can become aware and supportive of their children's mathematical interest, particularly through attendance at prize-giving sessions. Teachers become involved in the preparation of students, as members of problems committees, in organization and in invigilation. The contest can bring together and encourage teachers of similar interests and broaden their mathematical experience and perception. For academic mathematicians, there is the obvious contact with elementary and secondary teachers and students through cooperation and visits to schools.

Beyond the academic community, sponsors are a vehicle for broadening public interest. They have a direct involvement, and if they are substantial and conspicuous, they can generate publicity and attention from the media.

The main danger to be avoided is the discouragement of children through inappropriate questions and lack of preparation. Candidates should know in advance how the examination is to be run and what sort of questions are to be set. They should have the mathematical background necessary, some advice on sitting examinations and adequate psychological preparation.

Written competitions are not the only vehicle for arousing mathematical interest in the young. There can be active participation events, such as mathematics leagues, which embody team and cooperative activity. Other events would include competitions for displays and exposition, awards for teachers based on innovative ideas, and activities such as Maths Trails, which would involve the whole family.

Prizes, as was written earlier, should be numerous and should themselves be chosen so as further to popularize mathematics - books of the type we have described earlier would seem particularly appropriate.

14. RADIO

The influence of TV is now so great that the claims of radio as a medium are often overlooked. Yet, in some countries radio is still the major means of communication, in others the particular properties of radio are recognised and exploited alongside those of television. One strength of radio, for example, is that it forces the individual to visualise - it does not impose a particular visual image upon the listener. Clearly, this may make exposition more difficult, but handled with imagination it can make learning more creative and effective.

In many countries radio is still a major teaching medium, playing an important role in mathematics education yet outside the remit of the Leeds seminar.

Elsewhere, it is used to popularize (see, for example, the paper by Barbeau and the reference earlier under competitions). In particular, the Leeds seminar heard of interesting work being done on 'France-Culture', the French national, cultural channel. (Reciprocally, listeners to 'France-Culture' heard an hour-long programme on 31 October, 1989 about the Leeds meeting.) Two recent series seemed of special interest, one on the history of Bourbaki and the effects of that movement, the other a four-part series, 'Les délices des mathématiques', each part lasting 80 minutes. The guiding motto of the series was 'pas de cours, pas de discours, des témoignages' ('no lessons, no lectures, only testimonies'). The programmes took about a year to prepare and were presented and devised by mathematicians. The main focus was the world of research in (pure) mathematics, impact on daily life, human aspects and historical/philosophical origins. Those interviewed ranged from world famous mathematicians, eg. Cartan, Schwartz, Connes, Serre, to young students and scientists from different backgrounds, a philosopher, historians, a psychoanalyst, a musician, and secretaries of mathematics departments. No precise themes were pre-defined but the interviews ranged widely in an attempt to allow listeners to understand who mathematicians are, what they do, why, what the effects are, ...

Clearly, the number of listeners to this series would have been miniscule compared with those who viewed, say, the TV programmes on 'chaos' or 'Ramanujan', or who watch 'Square One' regularly. Yet this serves to highlight another key issue on popularization. Efforts directed at a large audience may, of necessity, be distorted and, to some degree, unrepresentative and misleading. This is a price that must be paid. The changes will be reduced if media people and mathematicians work together closely and with mutual respect. The former know how to attract and hold attention, the latter have to help more clearly identify and clarify the mathematical messages to be transmitted. Yet there have to be events which are directed more at minorities and which present a fuller picture. 'Les délices' would seem to have achieved its aim of setting mathematics in historical, epistemological, philosophical, scientific and social contexts: of showing mathematics to be part of contemporary culture. The programmes might have "pleased not the million; 'twas caviare to the general", but popularisation should include caviar as well as cream cakes!

15. OTHER MEDIA

The media we have briefly considered above were those to which most discussion time was allotted at Leeds. However, various other media were touched upon and reported. As mentioned earlier, lectures were given some consideration, yet the particular problems these set the speaker and ways in which these could be overcome were not identified in any systematic manner. The need to match approach to audience is a prime one, and this is well illustrated in Tony Shannon's paper. In general, one suspected that a guiding, pragmatic principle appeared to be - if one can keep an audience engaged and interested for 50 minutes or so by a talk which is seen to have links with mathematics, and which leaves the listeners more sympathetic to mathematics and mathematicians, one has succeeded. Other considerations, perhaps rightly, were considered secondary.

The view is often expressed that listeners to a lecture are, of necessity, 'passive'. This is, of course, nonsense. Yet, those attending a lecture have no control over the way in which they are intellectually active: the pace and emphases are set for them by the lecturer. It is in this respect that workshop activities have so much to offer. Workshops intended primarily for children are described in the paper by Hatch and Shiu. Mathematical camps are mentioned by others. At Leeds, Virginia Thompson, described how the Family Math project in the US had also used the workshop to involve parents in a given community. It was suggested that workshop activities might also be an excellent way of popularizing mathematics amongst retired persons - for then they would serve not only a mathematical purpose, but a wider, social one.

The general aim to involve parents and others in mathematical activities, whether through workshops, exhibitions, puzzle solving and competitions, was given repeated emphasis. Yet another way in which this had been done was through 'mathematical trails'. These guided tours of cities and sites, which draw the followers' attention to mathematical problems, results and artefacts to be found in the environment are now becoming more popular. Dudley Blane described certain Australian examples to the seminar and spoke of their success in capturing the interest of young and old - and not necessarily in separate groups.

This, again, leads to another apparent key principle relating to popularization: an activity that cuts across age divisions and can be shared by a family or several generations has particular value. For example, observers at exhibitions such as the Pop Maths Roadshow will readily note a difference in the nature of responses to exhibits between family groups and, say, classes of schoolchildren. Each generation looks at an object or activity in a different light, and the consequent exchange of views and experiences can be very enriching.

16. FINAL THOUGHTS

The Leeds seminar helped those attending realise the great number of ways in which attempts are being made to popularize mathematics. It is unlikely that anyone left Leeds without having been introduced to some new example of considerable value.

Yet, the meeting also left other impressions. It was not always the case that objectives had been clearly identified, or that there had been any serious attempts to see whether objectives, if any, were being met. (Here one notes the need for yet another column in the grid described in Section 5: '*our* motivation'. For we cannot have exactly similar objectives in mind when addressing different groups.) In the case of those using 'public media', such as TV and the press, then they were subject to normal evaluative procedures (if only the crude one of a viewer count). Yet, here, we had an example of evaluation of considerable significance: for regular viewers of 'Square One' had been found to do better at conventional, school mathematics. This can be seen as supporting the argument expressed elsewhere in this volume that the primary aim of popularization is not to teach mathematics but to change attitudes. Once attitudes have been changed, or even merely relaxed, then learning mathematics, appreciating the subject and what its practitioners do, etc, may be significantly affected.

Nevertheless, popularization in this sense is merely throwing out lifebelts to strugglers. Adverse attitudes towards mathematics are not congenital: they are built up through school and contact with elders. The latter influence will not be readily remedied nor will it be easy to change the former, but changed it must be. Some hints on how changes might be made are contained in this book, but as we wrote earlier there are deep problems in learning mathematics which cannot be eradicated - all we can do is provide motivation for grappling with them.

Once changes in attitudes are achieved there is still, however, a _need_ for popularization. Changes in attitude can only lead to a demand for information: popular lectures are, in the main, attended by those who _want_ to know; the France-Culture programmes were listened to by those who wished to understand better. Here we are concerned with mathematics which will not be examined, is not needed in order to proceed to the next course, and may well have no immediate relevance to employment. A change in attitudes will not, therefore, do away with the need for popularization; it would, however, make it much easier for mathematicians to meet its demands.

Mathematics in Different Cultures

REPORT OF THE WORKING GROUP

This group felt that although its brief might have appeared limited at first sight, it has important points to make to everyone concerned with popularization. The term 'culture' can, and should, be interpreted broadly in order for popularization to stand any chance of success.

The key aim of popularization is to overcome alienation. We identified power imbalance in society as one of the fundamental causes of alienation, with 'Western' Mathematics being seen to be a strong part of the 'educational' system helping to alienate various groups in different societies.

In some countries there are indigenous cultural groups as minorities (eg New Zealand, Australia, USA, Canada, Finland) and in the majority (eg South Africa) though in all those countries the dominant cultural group assumes Western Maths to be the only mathematics worth knowing.

In Africa and South America there are ex-colonial societies trying to identify their own view of mathematics, while in Europe, North America, and Australasia there are new immigrants feeling alienated from the 'resident' culture.

In all these situations it is as much the process of cultural alienation which needs to be overcome as the dominant mathematical view itself. This implies that the following points need particular consideration:-

1) 'Who does the popularizing?' is a key question. Basically 'we' can't do it for 'them', and we need to recognise the need to develop such notions as bilingual/bicultural units, family and community groups and leaders, indigenous leaders and popularizers.

2) Most popularization is carried out in the language of the dominant group and this issue needs addressing. Culture and language are intertwined, and language is for many the heart of their culture. 'Their' language expresses 'their' mathematics.

3) Everyone in Mathematics and Mathematics Education needs to be aware of the cultural nature of mathematics. Western Mathematics is a particular form of knowledge having a particular cultural history. This fact needs to inform all kinds of popularization.

4) Awareness is not enough though, and in the context of this seminar, legitimation is crucial - that is, popularization must legitimise mathematical ideas which are not in the dominant mainstream. It means legitimising other forms of mathematical knowledge and values, and it means legitimising the activities of those mathematicians who practise in other cultural groups.

5) There are appropriate and inappropriate ways to talk about knowledge and to use knowledge in different cultures. This demands sensitivity within the popularization process, encouraging again the need for other cultural representatives to be engaged in the process.

6) The early mathematical knowledge of the dominant group should not be ignored in any popularization, otherwise there is a danger of other

cultural knowledge being projected as primitive and inferior. In other words, <u>old</u> non-Western Mathematical ideas should not be contrasted with <u>new</u> Western ideas.

7) There are significantly different conceptual frameworks in different cultures and mathematical ideas will not necessarily be separated from other ideas in the ways that Western Mathematics is.

8) Care should be taken not to glorify, or make exotic, other peoples' culture. One may well be referring to a Westerner's <u>historical</u> version of that other culture which may not coincide with the other person's <u>present</u> views.

Finally we considered that ICMI has a key role to play in this area in the following ways:-

ICMI should encourage any attempts to popularize which are tackling this cultural dimension. It is a relatively unexplored area, it is extremely complex, and <u>urgent</u>. Cultures are fighting for survival.

ICMI should invite more involvement from alienated cultural groups in its activities.

Regional meetings would be particularly appropriate for addressing those issues, but the ICMEs are also extremely important events for increasing awareness and sensitivity.

However the most important role for ICMI is to <u>legitimise</u> other mathematical activities besides those which are identified with the dominant cultural group. As an international organisation, ICMI should have a truly multi-cultural perspective, and this perspective should influence all its activities, its publications and its structures.

A.J. Bishop (Group Reporter)

Mathematics for the Public

EDWARD J. BARBEAU

Department of Mathematics, University of Toronto, Toronto, Ontario, Canada M5S 1A1

"With some astonishment Hans discovered how different all things looked to his friend than to him. Nothing was abstract for Heilner, nothing he could not have imagined and coloured with his fantasy. When this was impossible he turned away, bored. Mathematics, as far as he was concerned, was a Sphinx charged with deceitful puzzles whose cold malicious gaze transfixed her victims, and he gave the monster a wide berth."
    Hermann Hesse, *Beneath the wheel* Bantham, 1970.

"Bridge is much more than a game of inference and logic than of mathematics."
    Steve Becker, *Contract bridge* Toronto Globe and Mail, February 25, 1989.

"Finally, we call attention to one additional aspect of the preceding analysis which may be of interest to teachers of mathematics. This is the fact that our result provides a handy counter-example to some of the stereo-types which non-mathematicians believe mathematics to be concerned with.

Most mathematicians at one time or another have probably found themselves in a position of trying to refute the notion that they are people with a "head for figures", or that they "know a lot of formulas". At such times it may be convenient to have an illustration at hand to show that mathematics need not be concerned with figures, either numerical or geometrical. For this purpose we recommend the statement and proof of our Theorem 1 [*There always exists a stable set of marriages*]. The argument is carried out not in mathematical symbols but in ordinary English; there are no obscure or technical terms. Knowledge of calculus is not presupposed. In fact, one hardly needs to know how to count. Yet any mathematician will immediately recognize the argument as mathematical, while people without mathematical training will probably find difficulty in following the argument, though not because of unfamiliarity with the subject matter.

What, then, to raise the old question once more, is mathematics? The answer, it appears, is that any argument which is carried out with sufficient precision is mathematical, and the reason that your friends and ours cannot understand mathematics is not because they have no head for figures, but because they are unable to achieve the degree of concentration required to follow a moderately involved sequence of inferences. This observation will hardly be news to those engaged in the teaching of mathematics, but it may not be so readily accepted by people outside of the profession. For them the foregoing may serve as a useful illustration."
> D. Gale & L.S. Shapley, *College admissions and the stability of marriage*. Amer. Math. Monthly 69 (1962), 9-15.

Probably no area of human activity is as afflicted as mathematics with a gap between the public perception of its nature and what its practitioners believe it to be. Contrary to the second of my quotes, inference and logic are as much part of mathematics as computation. But so also are experimentation and discovery, beauty and organization of ideas. Because of the esoteric nature of frontier research, mathematical expositors have a particularly difficult task in suitably conveying to the layman the beauty, fecundity and power of mathematics. I will present some ideas which I have used in various settings - on the radio, in talks and in a mathematical column for a magazine.

1. QUIRKS AND QUARKS

Quirks and Quarks is a science magazine radio show broadcast by the Canadian Broadcasting Corporation each Saturday from 12 noon to 1 o'clock for an

Mathematics for the Public 43

audience of about 400000. In each of about five minutes, host Jay Ingram interviews a scientist about current research. Although virtually every area of science is covered, little has been said about mathematics. Current developments being mostly inaccessible to a general audience, one has to resort to well known or historical material. For example, Len Berggren of Simon Fraser University presented capsule biographies of mediaeval Islamic mathematicians (his research area).

During the past year, I have presented six items on this show. In each case, they were discussed with the producer, Anita Gordon, before I was interviewed on tape by Jay Ingram in the role of a typical listener with a general intellectual interest but no special mathematical background. To preserve spontaneity, the tapes were broadcast unedited, one item every couple of months or so.

(a) *Primes* In this interview, Jay and I talked about prime numbers, long-standing conjectures (prime twins and Goldbach) and some recently discovered large primes.

(b) *Collatz sequence* I told Jay about sequences formed as follows: *begin with any positive integer; if it is odd, multiply by 3 and add 1; if it is even, divide by 2; iterate this process to generate the sequence.* Jay worked out an example beginning with 7, describing to the audience what he was doing. I then explained the notorious conjecture that, regardless of the starting number, the sequence eventually settled down to 1, 4, 2, 1, 4, 2, 1, ... The week after this was broadcast, a history professor mentioned that this conjecture led to some experimentation on his part.

(c) *Möbius strip* In this episode, we had in the studio a long strip of newspaper, some sticky tape and a pair of scissors. At my instruction, Jay made a Möbius strip and cut it down the middle. He conveyed to the listeners his astonishment at finding that it "stayed in one piece when divided". One indication of the attractiveness of this item was the participation of the technicians in the control room who performed their own experiment, which, Jay assured me, was quite rare. Shortly after this was broadcast, a letter from a schoolgirl asked for advice on a Möbius strip project.

(d) *Sum of cubes is square of sum* I went through with Jay the calculation that $(1+2+3+4)^2 = 1^3 + 2^3 + 3^3 + 4^3$ and explained that this was part of a more general result. We then discussed a way of constructing a set of numbers the sum of whose cubes is the square of the sum. To illustrate, pick any number, say 12. Write down its divisors and beside each divisor the number of its divisors:

Divisor of 12	Divisors of divisor	Number	Cube of number
1	1	1	1
2	1,2	2	8
3	1,3	2	8
4	1,2,4	3	27
6	1,2,3,6	4	64
12		6	216
	SUMS	18	324

Jay and I worked through this example, and he registered the appropriate amazement at the relationship between the two sums. Afterwards, I spoke to a person who claimed to be interested in the item, but got lost during the discussion of the example. It seems essential to have the listener following with pencil and paper.

(e) *The number 142857* The interest of this number, shown to me by my grandfather when I was a lad, lies in the fact that the first six multiples have the same six digits rearranged cyclically: 142857, 285714, 428571, 571428, 714285, 857142. The seventh multiple is 999999, and this gives the key to the phenomenon. Jay and I discussed the role of the prime 7 and the divisibility of a number with 6 = 7 - 1 nines by 7.

(f) *Perfect numbers* Here we talked about the definition of perfect numbers, how to find even ones and the open problem about the existence of odd ones.

I am not aware of any systematic analysis of the effects of these items. However, I have met several people from different backgrounds who heard them and seemed interested. Since the time constraint is severe and there is no access to visual illustration, there are few topics suitable for this format.

2. SOME OTHER IDEAS

My other experiences in conveying mathematics to a more general audience consist of sessions with high school students and their teachers, and the preparation of a puzzle column for the readers of the *University of Toronto Alumni Magazine* which appears quarterly. My intention is to encourage the listener or reader to experiment and to realize that there are logical and technical tools which can be brought to bear on a problem; it is not just a matter of luck or brilliance. Furthermore, I believe that much of the character of advanced mathematics can be authentically reflected in elementary materials. These aims can be more adequately fulfilled with a live audience and a half to full hour time slot, where some interaction and development of ideas is possible.

I have discovered a large reservoir of talent for the more instinctive side of mathematics, such as dissection of geometric figures, recognition of patterns and the solving of puzzles and games. On the other hand, it is difficult for many people to appreciate the economy and power of mathematical reasoning. For example, it is hard to grasp the generality of the pigeonhole principle; many people envisage only the possibility that each pigeonhole receives a letter before the doubling up occurs.

Here are some further examples:

(a) *Two-person zero-sum games of perfect information* Discussion of this topic can be built on experiences which are very common. Comparing the players in a game to the actors in an economic system can convince the reader or listener that there is a serious applied side, but it does not take long to get into questions of interest in their own right. The game of *Sim* is played as follows. Two players move alternately with pencil and paper on which there are initially six dots. A move consists of joining two dots not already joined (so that 15 is the maximum number of moves); the first player to complete a triangle with three edges he himself has drawn wins. This game involves some combinatorics (Ramsey's Theorem assures that no game ends in a draw), mathematical reformulation (reduce the game to one move on either side - *pick a strategy*) and a simple but sophisticated argument that the first player has a winning strategy. It is then revealed that the problem of actually finding a winning strategy is both difficult and recently solved.

Another two-person game of special interest has as alternate moves the choice of a number between 1 and 9 inclusive which has not already been selected. The first player to find among the numbers he has picked three which add up to 15 is the winner. This game is new to almost everyone, and it is by no means clear how best to proceed. However, the key is the magic square:

4	9	2
3	5	7
8	1	6.

Three numbers add up to 15 if and only if they are in the same row, the same column or the same diagonal. Thus, it is not hard to see that our game is isomorphic to noughts-and-crosses, with which almost everyone is familiar. In this way, one can convey the idea of mathematical structure and isomorphism, and the technique of transforming a given system into another which is more familiar or tractable. This central idea of mathematics is untouched by the school curriculum.

There are many other games which can be used to make a point in number theory, combinatorics or topology, such as Whytock's Game, Hex or Sprouts.

(b) *Using the imagination in topology and geometry* Two mathematical objects which appeal to many of the public are four-dimensional space and the Möbius strip. Starting with the idea of building up a square from a segment, and a cube from a square, we can then go on to the hypercube. By analogy, the audience can discover the number of vertices, edges, faces and cells and

Mathematics for the Public 47

devise strategies for visualization. In a similar way, the tetrahedron and octahedron can be generalized to four dimensions. A discussion of the Möbius strip leads on to the Klein bottle and the projective plane, and their representation by identifying sides of a rectangle. We can begin to develop the idea of a manifold and the way in which familiar objects can be used to help us think about abstract constructs.

(c) *Coding* A recent area of systematic mathematical development is ciphers. There are several topics which lend themselves to public exposition, but my favorite is the following. We have a number of symbols which we wish to represent by code words consisting of 0s and 1s (no blanks are allowed). We require not only that distinct symbols have separate code words, but also that the code word for one symbol is not the same as the beginning (prefix) of that for another symbol. The reason for this restriction is that at any point in the reception of the coded message, we should know whether or not we are in the middle of a code word without having to read ahead. If we try to specify the lengths of the code words in advance (symbol A is coded by a word of length a, symbol B by a word of length b, and so on), what conditions on the length will ensure that a suitable coding is possible? The answer is that the lengths $a, b, \ldots$ are characterized by the Kraft Inequality: $1/2^a + 1/2^b + \ldots \leq 1$. This is a bit of recent mathematics which can be made clear to an intelligent lay audience.

(d) *Getting the right perspective* A public used to the idea that mathematics is formal and manipulative is not prepared for the flexible thinking which will make a seemingly difficult problem almost trivial. Some of these problems are familiar, and the audience can be led to think about them in a more insightful way. Two such are the bird-and-train and the water-and-wine problems (Two trains initially a kilometer apart are travelling towards each other, each at 20 kilometers an hour; a bird, starting at one, flies back and forwards between them at a rate of 30 kilometers per hour. How far has the bird travelled when the trains meet? / You are given a litre of water and a litre of wine. One millilitre of water is transferred to the wine and the mixture stirred; one millilitre of this mixture is transferred to the water vessel. At the end, is there more water in the wine vessel than wine in the water vessel?) Another problem asks to maximize the area of an isosceles triangle with its equal sides given. If you take one of the equal sides as base and recall the half-base-times-height formula, the answer becomes inevitable.

(e) *Systematic analysis* There are many problems well established in the public domain which can serve to illustrate organized mathematical thinking. Most people attempt them relying on a mixture of guesswork, inspiration and luck, and it is a revelation that they can be approached in a systematic way, perhaps through the use of a tree diagram. Problems in this category include
- river-crossing problems such as the one about the wolf, goat and cabbage, or about the three jealous men and their wives;
- use of vessels with given fixed capacities to get an exact specified amount of water;
- equal-arms balance puzzles.

For example, you are given twelve billiard balls, eleven identical and the twelfth either heavier or lighter than each of the others. Using an equal-arm balance as seldom as possible, determine the odd ball and whether it is heavier or lighter. An analysis might note that there are 24 possibilities (ball 1 is heavy, ball 1 is light, etc.) and three possible outcomes of an experiment with the balance. After one use of the balance, the best that one can hope for is to have no more than eight (24 ÷ 3) outstanding possibilities, although there will be at least eight in the worst case. By means of such minimax reasoning, one can quickly realize that at least three applications of the balance may be required and hone in on a strategy which will never lead to any more than three.

(f) *Mathematical truth and power* I am indebted to John Conway for the following example. Start with an ordered pair of positive numbers. We extend this to a sequence by taking the last number obtained so far, adding 1 to it and dividing by its predecessor to get the next entry. An example is

$$5 \quad 7 \quad 8/5 \quad 13/35 \quad 6/7 \quad 5 \quad 7 \quad 8/5 \quad 13/35 \quad \ldots \ .$$

I invite each reader to do the same with any two starting numbers of his own

choosing; the periodicity is no accident. The discussion can deal with the following issues
- are all such sequences of period 5? how do we know for sure?
- if we try to avoid the use of vulgar fractions by using a calculator, are there any pitfalls? (the calculator truncates the decimal expansion and so may introduce a cumulative error)
- not only are these sequences dandy practice for children in manipulating fractions, there is no need to agree with an answer at the back of the book or to elicit a checkmark from a teacher in order to validate one's computations;
- its being impossible to establish the generality of the periodicity by looking at ever more examples, one introduces variables as stand-ins for numbers and imputes to them operational rules satisfied by numbers; thus, algebra appears as part of a proof technique;
- the algebra becomes unmanageable unless one simplifies the fractions obtained (at one point, by performing a factorization); thus, technical manipulations are important to gain access to information which might otherwise be hidden.

(g) *Experimental mathematics* Pattern recognition is one area which seems to work well with a lay audience; its open-endedness invites participation from a variety of people. There are a number of relations involving squares, such as the pythagorean equation, which are ideally suited for demonstrating the art of experimentation and conjecture. A sequence which is almost as good as the Fibonacci sequence for its richness is this one:

$$1 \quad 6 \quad 35 \quad 204 \quad 1189 \quad 6930 \quad 40391 \quad \ldots \,.$$

Taking sums and differences of consecutive terms allows us to generate close-to-isosceles pythagorean triples; for example,

$$6 + 35 = 41 = 20 + 21; \quad 35 - 6 = 29 \quad \text{and} \quad 20^2 + 21^2 = 29^2 \,.$$

Since $1 = 1 \times 1$, $6 = 2 \times 3$, $35 = 5 \times 7, 204 = 12 \times 17, \ldots$, one can broaden the discussion to approximating the square root of 2 and finding solutions if

Pell's equation $x^2 - 2y^2 = \pm 1$. With such examples the fecundity which fascinates mathematicians about their field can be illustrated.

(h) *Mental arithmetic* There is an Indian lady who travels from city to city giving demonstrations of her rapid calculating prowess. She begins her routine by finding cube roots of numbers of up to ten digits in her head. The skill of finding the cube root of a number of no more than nine digits is not too hard to teach. Knowledge that the given number is a perfect cube is a considerable start, for we have a lot of information from which to find only three digits. The last digit of the root is immediate; the first can be quickly determined by anyone who has memorized the cubes of the nine nonzero digits. The middle digits can be found from a combination of casting out threes and estimation. If an audience can be encouraged to look at number properties with a discriminating eye, it can master other mental arithmetic stunts with confidence and satisfaction.

(i) *Geometry* There is a wealth of material here, depending on whether one is dealing with dedicated amateurs who recall much of their school Euclidean geometry or with those who have the background only for basic geometric reasoning. I have found the columns in *Scientific American* by Martin Gardner a good source of material in this category.

These are just a few of the ways in which we can counter the abstruseness of most frontier mathematical research and yet convey some of the essence of mathematical work. We must develop material which require the technical background of a typical thirteen-year-old and can illustrate similarity of structure (isomorphism), use of symbolism and expression of ideas, algorithms and schemata, and the power of logical reasoning. We need to analyze everyday experiences and analyze them for their ability to give an authentic picture of mathematical values and activities. Finally, we have to discuss our experiences among one another so that our contacts with the public can be refined and improved.

Making a Mathematical Exhibiton

RONNIE BROWN AND TIM PORTER

School of Mathematics, University of Wales, Bangor, Gwynedd LL57 1UT, U.K.

1. INTRODUCTION.

For the last three years, a team in the School of Mathematics at the University of Wales, Bangor, has been designing the exhibition Mathematics and Knots, which was exhibited in the Pop Maths Roadshow in Leeds, 1989, concurrently with the ICMI Seminar. In this paper we explain what we were attempting to achieve, and the problems we had in getting to this stage.

We do not claim to have achieved all our aims, or to have reached a final version. The exhibition will be useful if it is enjoyed by the public and by mathematicians. We hope it will also stimulate others to think about the problems of exhibition design in mathematics and to encourage them to prepare for themselves presentations of mathematics in a variety of topics and media.

2. HOW IT BEGAN

Our involvement in making a mathematical exhibition came about in the following way.

One of us (R.B.) was invited to give a Popular Lecture on knots, one of two lectures in an evening, for June, 1984. It seemed a good idea to have material to display in the foyer for people to view when they arrived and in the coffee interval. Coloured enlargements were made of slides of knots in art and in history, and also a few of the overhead transparencies used in the lecture were enlarged to A3 size. All this material, and some models of knots made of copper tubing, was rather randomly distributed over some display boards.

During the next year we accumulated more material, which was used successfully on several occassions. In the Summer of 1986, we obtained limited sponsorship from Anglesey Aluminium to develop the exhibition in a more professional way, by bringing in a designer. We also set up a design team of R.B. (Chairman), T.P. and Nick Gilbert (now at Queen Mary College, London). It was the continual development of ideas by this team, with a full and frank range of suggestions and criticisms, from within the team, from our designers, and from others who saw versions as they developed, which has led to the exhibition as it now exists.

Versions and parts of the exhibition were shown at various venues between 1986 and 1988. In particular, in 1988 it was used at the Royal Institution to accompany a Discussion by Sir Michael Atiyah on "The geometry of knots". On the last occasion there was also an exhibition of some John Robinson sculptures, of Celtic Interlacing by Lady Wilson from the British Museum, and of knots by Geoffrey Budworth of the International Guild of Knot Tyers.

3. MAJOR PROBLEMS
We, and our designer, had not realised the extent of the job. The design work has already extended over almost three years, and has involved many drafts. The reasons for this were of several kinds.

3.1. Novel Aims
We set ourselves some novel aims for the exhibition (see section 3). In so doing, we were setting out into uncharted territory, and had to learn as we went along.

3.2. Lack of design experience
Initially we knew not even the elements of the mechanics of producing an exhibition, but our aims required a close marriage of content and medium. So we had to learn some of the problems and techniques as we went along. We were not able to employ the designer for the continued help we needed because of the lack of funds.

3.3. Insufficient funds
In order to finance the exhibition further, we wrote to a number of firms and organisations, and obtained a lot of help. A list is given at the end of this article. In 1987 the Committee for the Public Understanding of Science was set up by the UK Government.. We made an application to them and were fortunate to obtain a grant of £2,000. Without this total sum of about £3,860 the exhibition would not have been able to be produced.

To produce anything to a professional standard is expensive. An estimate we had recently from a member of a Government agency was that a 30 square metre exhibit had cost £20,000, of which half was for design and half for construction. In fact our designer gave us a lot of help for very little financial reward. He suggested the use of polystyrene boards with aluminium surround for durability, lightness, and transportability. He did a lot of initial printing layout. However it gradually became apparent that his initial layout of two columns per board was not appropriate for the exhibition, where the graphics had to predominate. The layout was considerably revised with the help of another designer. Finally, on the advice of the head of a local Arts course, we paid for the layout to be redesigned yet again using a grid approach, in which all modules of text and graphics fit into a centimetre square grid, and for all the knots to be redrawn to have a three dimensional effect and so as to fit into the modular layout.

A severe problem is printing costs. To produce an A2 board one first produces an A4 board with the text and graphics. This is photographed to an A2 negative from which the print is made. The enlargement from A4 to A2 (four times the area) means that the typesetting has to be with 1200 dots per square inch: the usual output of a standard office laser printer is quite unsuitable for this job. Commercial computer typesetting is expensive. The only way of proceeding within the budget then available was to use facilities in the University system.

At that time, the only system available was the Oxford Lasercomp system. This is an excellent system for producing a book in one or another standard text format. It also has a good variety of fonts, though not as many as commercial printers. It works however on what is called a "markup" system, where all control is by inserted commands in the text. This has the advantage of ease of sending files electronically. At first we simply sent down the text, so that the layout was achieved by pasting. Then we put the commands into the text to achieve the desired layout. However, the disadvantages of a markup system for complicated page layouts are great, since you do not see whether or not the layout commands have been correct until the proofs return days or, at difficult times, a week or so later. Further, the combination of graphics and text can only be produced by photographing large drawings to a small size for pasting on the A4 page, so that the enlargement produces a good quality product. Thus the design has got to be completely decided, and there is little room for experiment.

In the summer of 1988, we installed a Desktop publishing system (in fact Aldus Pagemaker©) with a scanner and Postscript printer. This allowed the drawings to be scanned and positioned within the text, with tight control of spacing and layout. Experimentation and modification was now easy. The initial work here was carried out over two years by three students on a course funded by the European Social Fund, and the final layout was done by T.P. Without this funding and equipment, the current production would not have been feasible. The final output was typeset at the University of London Computer Centre.

In each case, the techniques had to be learnt from scratch.

3.4. Stand-alone exhibition
The key difference from the design viewpoint between the foyer exhibition and our new aims was that we intended a travelling exhibition independent of any lecture. This meant that the boards had to be self-explanatory, and not just titillation for an explanation to be obtained later in the lecture. So we had to decide what story we wanted to tell. It was

at this stage that we began to formulate and clarify our aims, while we were designing the text and graphics for each board.

3.5. Visual impact
The visual impact in an exhibition has to dominate. Each board has to tell a story, but the story has to be told mainly through the eye rather than through the text. So the exhibition format is one of the hardest to get right for the conveying of ideas, rather than simply the presentation of images.

3.6. Use of language
It is very easy for a mathematician to use words and phrases which mean something to him but which convey nothing to the general public. As an example, we found ourselves using the phrase "uniquely up to order". In fact, this is a sophisticated idea, which we eventually conveyed by specific examples with numbers. The aim was to use a simple and clear language. The stripping of inessentials and unclear language was a part of the design process. On the other hand, we also found that some ideas needed a more leisurely exposition than we gave them at first. Each such change involved also a change of layout.

4. AIMS
The decision as to the aims of the exhibition was crucial. The issues involved are part of "exhibitology",(L.Brown 1988). This term has been coined by Len Brown (R.B.'s brother). He learnt the basics of this study in his period as Head of the Engineering Section of the Science Centre at Toronto, and suggested the following simple example as illustration of some basic principles. Suppose that you wish to produce an exhibition on "Bridges". There is certainly a lot of material available. However a decision has first to be made on the point of view to be taken. Is the exhibition to be about transport? structure? geography? history? rivers? trade? Each of these themes could lead to an acceptable exhibition, the various exhibitions would have much common material, but in the end each would be telling a different story, and give the visitor a different impression of the subject of Bridges. In our case, we had to decide what impression of mathematics we were intending to convey, and then seek to find the means to do so.

The aims we set for our exhibition fell into two kinds: structure, and content.

4.1. Structure
We agreed that the exhibition should be:
    a) self contained
    b) easily transportable
    c) reasonably cheap to produce
    d) reproducible in several copies
    e) able to be set up and managed without continual supervision .

4.2. Content
We agreed that the exhibition should:
a) suggest that the making of mathematics is a natural human activity, part and parcel of the usual methods by which man has explored, discovered, and understood the world
b) present each item with a purpose and context, and not just because it was something that could be shown or demonstrated
c) convey an impression of some of the key methods by which mathematics works
d) show mathematics in the context of history, art, technology and other applications.

5. CONSEQUENCES OF THE AGREED AIMS

5.1. Structural
Our requirements tended to rule out hands-on material, at least initially. Such material is expensive to produce and maintain; if it does not work it is worse than useless; it can be stolen, and indeed this is more likely the more attractive the material. In any case hands-on material can also suffer from being a *gizmo*, designed because it is hands-on rather than to make a point which illuminates the themes of the exhibition; nice to play with, but superficial. The participant is expected to exclaim "Wow!", but there is still a question as to what he or she has learnt. Of course this tension between the requirements of entertainment and arousing interest, and the requirements of instruction and information, is basic to the whole activity of exhibition design.

Our requirements also meant that we were initially intending a static exhibition: something to be looked at, and enjoyed, but not involving an activity. Once the structure and the content had been decided, it would still be possible to design hands-on or animated material which would advance our overall mathematical aims and which could be used as occasion demanded and allowed.

5.2. Content
It was in terms of content that we felt we were taking the more radical line, and the various features of these aims deserve separate paragraphs and discussions.

Mathematical content
5.2.1. Mathematics and discovery
The novelty and excitement of mathematics is conveyed by some of the major mathematical exhibitions (Horizons Mathématiques in Paris and also as a travelling exhibition, and the Mathematika at the Boston Science Centre). However we decided that it was necessary to analyse the basis for mathematical novelty and excitement, in order to clarify what we were intending to present.

We felt that this excitement and interest comes from the vision of new relations, and new kinds of order or patterns. It is not the whiz! bang! excitement of the amusement arcade. For example, it is extraordinary that the number p is involved not just with the ratio of the circumference of a circle to its diameter, but also with the description of population distribution. It is extraordinary that ,whereas we think a negative number cannot have a square root, such does exist if we allow a new kind of number, and even more extraordinary that these new numbers should have applications to the study of prime numbers, to the design of electronic circuits, to cosmology, and to the study of elementary particles.

An exhibition should convey some flavour of the real achievements of mathematics. If instead it simply presents an assortment of, for example, strange polyhedra, and states that these are the wonderful things mathematicians study, then it will be very easy for the public to be convinced that mathematics is hard, or weird, or both. Each exhibit should have a mathematical point and should explain its relations with other parts of mathematics and with other disciplines.

In the case of this exhibition, we felt the most surprising idea that could be conveyed in a way related to everyday experience was the analogy between knots and numbers revealed by the notion of a prime knot. It is for this reason, as well as for the needs of exposition, that we devote more than one board to this topic.

Surprising applications are also important for conveying some of the excitement of mathematics. In this case, we stress recent applications, such as to knotted orbits in weather systems, and to knotted DNA.

Key aims of mathematics are to show new perspectives, views, and order in what seems initially a tangle of unanalysable phenomena. This is one impression of mathematics that we wish to convey to the viewer.

5.2.2. Normality of mathematical methods
Here is where we feel we are really breaking new ground. Mathematics lacks an adequate discussion of methodology. A majority of students of mathematics do not know what they are doing or why they are doing it; they know only that they have to learn how to do certain things. Very few university courses attempt to explain the reasons for the development of a particular piece of mathematics, in some cases because the teachers are unaware of these reasons. However, experience shows that an analysis of the particular methods used in mathematics, and a relating of them to standard methods by which we explore and manage the world, is welcomed with relief by school teachers and students.

It has been said that the difference between a professional and an amateur is that an amateur can do things, in many cases as well as a professional, but a professional also knows how he or she does things. It is this knowledge, based on tradition, experience, perception, judgment and analysis, which gives the professional the confidence to produce work on demand and to certain standards.

Of course, in this exhibition we cannot hope to convey the whole gamut of mathematical ways of working. We are not interested in conveying technique. What we want to express is the mathematical equivalent of musicality - perhaps we should call it mathematicality? This is a horrible word, but its derivation should at least convey what is intended.

Often, mathematics is presented as a completed body of knowledge, whose development has been unrelated to the activities of human beings. The questions which motivated the whole theory in the first place are in teaching often simply omitted, and students and pupils are asked to appreciate the methods and the theory without context, without relevance to other mathematical or scientific activity, one might even say, without meaning. For example, how many books on group theory are there which mention the range of applications of group theory, from crystallography to modern physics, and which show how the exposition given fits into the wide mathematical and scientific context? The dehumanising of the presentation of mathematics has gone very far.

Our aim was to use the theory of knots to illustrate some of these basic methods of mathematics. Our listing and analysis of these methods carries no claim to finality. However such a listing is useful as a systematisation, and, more crucially, as a way of relating these mathematical methods to standard methods of exploration and analysis. Thus we illustrate the claim that the peculiarity of mathematics lies not so much in its methods, but in the material and the objects with which it deals.

5.2.3. The context of history and art

Our aim here was to remind the viewer of how knots have enormous richness and importance in the history of man. It has even been suggested that the Stone Age should be called "The Age of String". We were fortunate to have been told by Joan Birman of the oldest known knot: the Antrea net, dated 7,200BC, from the Helsinki National Museum. The net, found in a peat bog, was 30 m. by 1.5 m. with a 6 cm. mesh. It had stone sinkers and bark floats, and was made of willow bark. The knot used then is still used today.(See the paper *The net discovery of Antrea* by J.-P. Taavitsainen and M.Huurre, National Board of Antiquities and Historical Monuments, Helsinki.)

One can only speculate on the social organisation and lives of the people who constructed this net, and on the length of time such a technological achievement took to evolve. One can also sense that the early understanding of the form of knots, and its link with survival, is an expression of an early but by no means primitive geometrical feeling, an understanding that the form not only can be so but has to be so, by virtue of its logic.

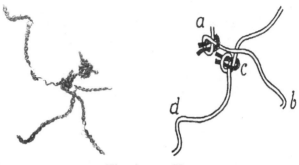

The Antrea Knot

We had planned a series of boards on knots in history, practice and art. However most of the design effort went into the mathematical boards and we felt that a full range of pictures was best presented as a slide show or video. In the event the video could not be produced to the necessary standard in the time left to us and so was omitted from the exhibition. One result, though, of this interest in the relations of Mathematics to Art was the exhibition of John Robinson sculptures at the Pop Maths Road Show (Robinson 1989). The practical side of knotting was represented by a large rope knot overhanging the exhibition, and two cases of knot work, all prepared by the International Guild of Knot Tyers.

6. Some mathematical methodology

As the structure of the exhibition developed, we arrived at the following list of some basic methods in mathematics which we were able to present:

    a) Representation
    b) Classification
    c) Invariants
    d) Analogy
    e) Decomposition into simple elements
    f) Applications

We now analyse these in detail.

6.1 Representation

In the case of knots, we have to show them and present them. This is usually accomplished by a planar diagram of the knot. But the difficulty of this step should not be underestimated. For example, we have found it helps children to make the jump from the diagram to the knot and back again by giving them pieces of string and asking them to make the knots according to specific diagrams. In the exhibition, we use shaded drawings to give a 3-dimensional effect.

A Mathematical Exhibition

6.2 Classification

Making lists is a basic human activity. However, in view of the complexities of the world, you cannot make a list of everything, at least not if you are to lead a sensible life. You may recall the Memory Man described by Luria (1968), who was unable to forget anything, and consequently was not able to lead a normal life. One of the diagnostic features for autistic children is that they remember nonsense patterns as easily as other patterns (Wing 1976).

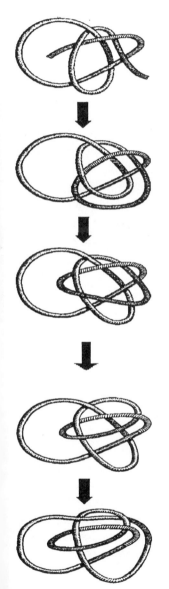

In making lists we impose or find order: we classify. For example, a zoologist does not list all the animals in a game reserve, he lists antelopes, elephants, lions and so on. In order to do so, he needs criteria for saying that two animals are the same. In mathematics the notion of equality, or, in more precise mathematical terms, equivalence, is basic.

Knot theory presents interesting mathematical points in this area. Firstly, "When are two knots the same?" is a non trivial question. Secondly, to make an initial list of the first elements of an infinite family involves some classification into the "simplest" elements. Thirdly, the presentation of even such a simple list is likely to suggest the need for some further order and classification.

We address the first question by showing in our boards how diagrams of knots can be transformed, without changing the knot. In particular we showed a sequence of pictures transforming a picture of a "standard" Bowline knot to one having only six crossings. (We have included a copy of this sequence on this page.)

This is one area where computer animated graphics could greatly improve the presentation. However, we are limited by the cost of producing the graphics, and the cost of presenting the graphics at an exhibition display. The latter is not so hard to overcome, since a video could be made and easily displayed, for example at most schools. However, it is more important for us to show the context in which any computer graphics would run, as this would dictate what graphics should be produced in order to support the overall themes.

One of the goals that we set ourselves was to explain the meaning of a list of knots. Although such a list is apparently simple, the explanation involves the following ideas: *when are two knots the same; crossing number; mirror images; the arithmetic of knots; prime knots.* These themes give interconnections between the different boards.

6.3 Invariants

Classification of knots involves two aspects: when are two knots the same? and when are they not the same? The first usually involves transforming one diagram of the knot into another. The second involves the more subtle point of deciding when such a transformation is not possible. Such a decision involves the notion of *invariants*.

We deal with four invariants in our presentation: *crossing number; unknotting or Gordian number; colouring number; bridge number.* We also mention briefly the new knot polynomials which enable one to distinguish easily between a trefoil and its mirror image. The advantage of the four invariants we deal with in detail is that they can be easily presented at this level, and that they suggest many detailed exercises and examples which people can try for themselves.

The further point made by the discussion of invariants is that we do not claim to give a complete set of invariants, that is, we do not have some method of distinguishing all possible knots. Thus many problems remain in the theory, and this again is a point which is easily conveyed. We do want the reader to see that mathematics is, and will continue to be, an open-ended activity.

6.4 Analogy

In order to explain our list of knots, we have to describe the notion of *prime knot*. This is done by explaining a basic composition of knots.

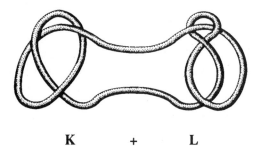

K   +   L

There was a decision to be made here: whether to call the composition of knots *addition* or *multiplication*. The literature uses both terms. We chose the term addition for two reasons. One was that the notation 0 for the unknot is more descriptive than the notation 1.

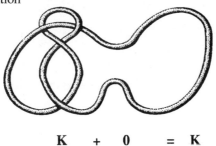

K + 0 = K

The more important reason was to emphasise that the analogy is not between things but between the way things behave, between their relationships. For this it is helpful to have a different notation for the two operations which are being taken as analogous. Instead of making an analogy between two multiplications, we make an analogy between an addition and a multiplication. This, we hope, is more striking and also illustrates a general point, that such analogies may be available in other situations.

Although we use the term "The arithmetic of knots", we also use algebraic notation and emphasise laws:

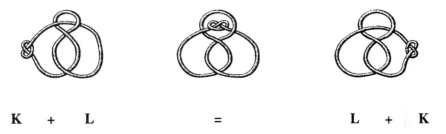

K + L   =   L + K

This makes the point that it as the study of analogies that algebra obtains its generality and power. By using the commonplace word "analogy", we aim to demystify, and to show that one aspect of the method of algebra is as a standard process by which we understand and try to make order of the world. The other point to be made is of course the excitement of an unexpected analogy, of "That reminds me of ...!".

We are also able to present the deep fact that *knots have a decomposition into a sum of prime knots, and this decomposition is unique up to order.* The appreciation of this led one small boy after the Mermaid Molecule Discussion by R.B. to ask: "Are there infinitely many prime knots?". It so happened that the lecturer had not previously formalised the question for himself, so had really to think in order to be clear that all torus knots are prime, and that there are an infinite number of them. However the proof that torus knots are prime is not so easy. It is good to have something to state which is comprehensible and believable, but which it is not at all clear how it might be proved. This is one of the great advantages of knot theory for expositions at this level.

6.5 Decomposition into simple elements

Decomposition into simple elements is a basic process in mathematics, or indeed wherever one deals with complicated matters. In knot theory the process crops up in a variety of guises.

a) We have already mentioned the prime decomposition of knots. Here the prime knots are the simple elements and the fact that any knot can be expressed uniquely (up to order) as a sum of prime knots is clearly an important fact about knots as well as simplifying the preparation of a table of knots.

b) The process of transforming one diagram of a knot into another may be quite complicated. It is therefore of interest that such a complex process can be resolved into a sequence of simple moves, the Reidemeister moves:

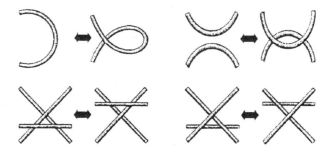

We illustrate this in the process of changing the Bowline (shown earlier in this article), and also in illustrating why the colourability of a knot is an invariant.

6.6 Applications

The exhibition starts with a picture of the sculpture Rhythm of Life, by John Robinson, (1989). It ends with an indication of some applications, including how knotting is involved in DNA, one of the building blocks of life itself.

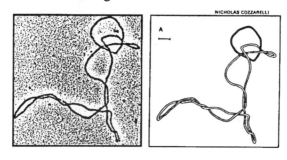

This section on applications could be expanded, and made more vivid in a variety of ways, given funding, space, and so on. But all we hope to do in this exhibition is to catch the imagination of some people. The exhibition is intended to be small scale, and unambitious in its use of technique. Indeed, this is imposed on us by the criteria which we outlined at the beginning. Such limitations are still compatible with high expository aims.

7. CONCLUSION

We believe that if the exhibition is successful on the terms initially laid out, then it should be possible to build out from it as wider funding and staffing, and more ideas, become available. The honing of ideas and presentations, the discarding and developing of innumerable drafts, the criticism and comments from many, all have been valuable in clarifying our aims and our methods. In particular, we give our thanks to the long suffering designers, Robert Williams, Jill Evans, and John Round, who have been involved at various stages.

The existence of various drafts of the exhibition has enabled Heather McLeay to start designing a set of worksheets for young pupils (McLeay 1988). Drafts of these worksheets have been used successfully at Royal Institution Mathematics Masterclasses in 1988 and 1989 at Bangor and Cambridge. We have also prepared a brochure based on the exhibition,(Exhibition Group 1989).

REFERENCES

Brown, L.M., (1988), Private conversation.

Brown,R., (1989), Conversations with John Robinson, in Robinson (1989) below.

Exhibition group,(1989), *Mathematics and Knots*, Mathematics and Knots, University of Wales, Bangor .

Luria,A.R., (1968), *The mind of a mnemonist*, trans. by Lynn Solotaroff, Basic Books, New York.

McLeay, Heather, (1988), Worksheets on knots , (in preparation), UCNW .

Robinson, J., (1989), *Symbolism: sculptures and tapestries*, (catalogue for the exhibition at the Pop Maths Road Show and the International Congress of Mathematical Instruction, Leeds, 1989), Mathematics and Knots, University of Wales, Bangor.

Wing, Lorna, (1976), *Early childhood autism: clinical, educational, and social aspects*, Pergamon Press, Oxford .

Design team
Ronnie Brown, Nick Gilbert, Tim Porter

Sponsors
Anglesey Aluminium; University of Wales, Bangor; British Gas; London Mathematical Society; Committee for the Public Understanding of Science; Midland Bank; Ferranti; Pilkingtons; Bridon P.L.C.

Assistance
John Derrick, Jill Evans, Gwen Gardner, Tony Jones, Jose-Maria Montesinos, Hugh Morton, John Round, Philip Steele, Robert Williams.

The Role of Mathematical Competitions in the Popularization of Mathematics in Czechoslovakia

VLADIMIR BURJAN

Gymnazium A.Markusa
ul.Cervenej Armady 18
811 09 Bratislava
Czechoslovakia

ANTONIN VRBA

Vyzkumny ustav pedagogicky
Mikulandska 5
116 56 Praha
Czechoslovakia

1. POPULARIZATION OF MATHEMATICS IN CZECHOSLOVAKIA

The situation concerning the popularization of mathematics (PoM) probably differs substantially from country to country. Trying to characterize the situation in Czechoslovakia, the following features should be pointed out:

(1) The main sections of public targetted are children, pupils and students. Almost no efforts are made to target another groups of citizens: adults, workers, professional scientists and specialists in other branches.

(2) The schools, unfortunately, play rather a negative role: maths is often presented as a rigid collection of formal theorems, formulae and proofs with no history, no development and no connection with the pupils' everyday-life experience. It often evokes feelings of tediousness, uselessness or fatal incapacity to understand it.

(3) On the other hand, a wide network of out-of-class activities has been developed, the majority of which are organized by small groups of enthusiastic mathematicians, mathematics teachers and university students, often only with a meagre financial support from local authorities. Nevertheless, this approach to popularization prove to be very efficient and vital. Here the main work in PoM in Czechoslovakia is done.

(4) The mathematicians themselves do not care enough about PoM. This results in relatively small numbers wishing to study mathematics or to become maths teachers. Therefore the quality of maths teachers and that of teaching mathematics is lower => the image of mathematics is poor => there is a shortage of those wishing to teach ... and we fall into the well-known cycle which was discussed (but not solved) in several sections at ICME-6 in Budapest.

(5) The contribution of such media as TV, film, radio and newspapers is minimal; their possibilities remain fairly unexploited.

With respect to the situation described above, it is obvious, that if we want to contribute positively to the work of this seminar, we have to focus on the out-of-class activities. Among them mathematical competitions play a substantial role.

2. MATHEMATICAL COMPETITIONS (some general notes)

Mathematical competitions (MC) are without doubt one of the most efficient means of PoM among children, pupils and students. Thanks to children's natural tendency to compete, MC have a strong motivational power. But motivation (i.e. influencing the children's emotions) is not the only purpose of MC. These may serve as well as:

- encouragement for a deeper study of mathematics and/or reading mathematical (popular) literature

- an opportunity to identify (and foster) gifted pupils

- one of the criteria of assessment of maths teachers (This is an often discussed problem and it requires caution) etc.

As we are discussing MC in the framework of PoM, we shouldn't restrict ourselves to competitions for gifted pupils. Thus, a natural question arises:

What should be the differences between an *"exclusive" competition* (designed for gifted pupils and aiming at fostering them) and a *"popular" competition* (targetting wider sections of children and aiming at raising their interest in maths) ?

One of the factors making a competition more or less popular and accessible is the sort of problems proposed. Obviously, the greater the distance between the type of contest problems and the exercises in the text-books, the harder it is for an average pupil to be successful. Thus, e.g. the International Mathematical Olympiads (IMO) and probably all national rounds of olympiads are quite exclusive and require hard training together with a great portion of giftedness. It is often stated that the olympiads (and similar problem-solving competitions) contribute considerably to the PoM. As regards the lower rounds, it is undoubtedly true. But the national and international rounds can hardly be viewed as a means of popularization because all their participants must already have a very close affinity to mathematics.

Another difference between the two sorts of competitions should be the level of "formal mathematical culture" required from the contestants. At IMOs the students are expected to express their ideas in a sound mathematical language using conventional symbols and supporting each step in the solution by precise argumentation. On the other hand, when running a popular competition we have to show much tolerance in this respect and find completely different criteria of assessment.

When speaking about olympiads, we cannot resist to call the attention of the reader to one problem concerning this type of problem solving contest. Namely, that there are substantial differences between solving olympiad problems and the scientific work of an mathematician (both pure and applied - if we accept this distinction). Here is a brief comparison of these two kinds of activities:

THE OLYMPIAD CONTESTANT:	THE MATHEMATICIAN:
receives the problems formulated	searches for problems himself
has to solve the problems within an allotted time	usually has not to care about time
often merely has to choose the appropriate method from a stable repertoire.	has to create completely new methods. These are often more interesting than the results
almost never creates new concepts	introduces new concepts (which often represent the most valuable achievements)
terminates by finding any solution of the given problem	the solution of every problem raises further questions
as a rule is not interested either in the origin of the proposed problems nor in their applications in practice or within the framework of a mathematical theory. (Such applications often even don't exist, since the problems are quite artificial)	studies each problem in the framework of a complex theory and investigates its applications within this theory and in practice

It is by no means our intention to criticize olympiads or to call their importance in question. We just want to point out the danger of misinformation of successful olympiad participants: a first prize winner at IMO may have a quite distorted idea of what it means "to do mathematics" or "to be a mathematician". This misunderstanding could perhaps be avoided by:

- presenting competitions merely as a connecting link between school mathematics and "real" mathematics. They must remain a means and mustn't become a goal.

- creating other types of MC, the nature of which would more closely resemble mathematics done by mathematicians. The first step in this direction are various types of project competitions being organized in several countries (including Czechoslovakia - see later). In our opinion, these popularize "doing mathematics" much more than the classical olympiad-type MC.

Another question arising when discussing the role of MC as a means of PoM is the following:

By what means can it be ensured that the experience of taking part in a (popular) MC is enjoyable for the pupils ?

Let us list some of the possibilities:

- a wide range of various competitions should be created to cover different age groups and different types of children's interests (problem solving/project, multiple-choice/essay type, team/individual, relay,...) so that every pupil can choose the competition best attuned to his/her character.

- the rules of an MC have to be interesting, i.e. they have to allow the contestants to choose various strategies how to proceed.

- the problems proposed should be attractive and not too hard.

- very much can be done by an appropriate choice of contest venue, by combining the competition with other interesting mathematical events, by giving attractive prizes and by providing the competition with adequate publicity among schools, local educational authorities and media.

Speaking too generally, one always risks a decline to banalities. To avoid this, let us consider the Czechoslovak MC in more detail.

3. MATHEMATICAL COMPETITIONS IN CZECHOSLOVAKIA.

In Czechoslovakia, MC have a relatively long tradition. The eldest one - MATHEMATICAL OLYMPIAD - started 38 years ago and since then it has made a remarkable progress. Its present structure is quite similar to that of other national MOs, therefore we shall restrict ourselves to a brief description.

- It is divided into 8 separate categories (according to the age of the pupils) and it covers all age groups from the 4-th grade (age 10) of our basic school up to the 4-th grade (age 18) of our secondary school. Besides, a special category "P" exists (at the secondary level only). The problems proposed concern programming and construction of algorithms.

- In each category the contestants pass through a hierarchy of rounds starting with a school round and finishing with regional, county or even national rounds (according to the respective category). In each round the pupils have to solve 3-4 problems within 3-4 hours. The national round of the highest category "A" lasts for two days and its participants have to solve 2×3 problems within $2 \times 4,5$ hours.

- The problems proposed do not go beyond the curriculum but they are quite different from the standard school exercises. The written solutions must contain a detailed argumentation and explanation of each step. Thus good results can be reached only by systematic training.

- For the best participants in each category special meetings (lasting for 1-2 weeks) with scientific programmes are being organized.

Using the terminology introduced above we might say that the lower categories of our MO (at the basic level, age 10-14) are a popular competition, while the higher categories (secondary level, age 15-18) represent a typical exclusive competition. In Czechoslovakia we have a wide net of special secondary classes for mathematically gifted pupils. (The first author is a maths teacher in such special classes. But it is not possible to discuss them here in more detail.) As a rule, the vast majority of the most successful participants of the final rounds of our MO come from these classes. This only confirms the exclusivity of the MO.

Here is a small sample of problems from various categories of our MO:

(1) (basic school, 4-th grade, age 10):
Arrange the digits 0,2,5,7,8,9 into the least possible even 6-digit number. (No digit may be used more than once.)

(2) (basic school, 6-th grade, age 12):
Patrick filled a squared table 2×2 with numbers 1,2,3,4 in such a way, that all sums of rows and columns were prime. Then he tried to fill a square table 3×3 with numbers 1,2,3,4,5,6,7,8,9 in the same way. Explain, why he must fail.

(3) (basic school, 8-th grade, age 14):
For which natural numbers n is there an n-gon with the following two properties:

I. all its sides are of the same length
II. every two adjoining sides are perpendicular to each other.

(4) (secondary school, 2-nd grade, age 16):
Find the greatest natural number n which cannot be expressed as a sum of two natural numbers, each with the sum of its digits greater than 9.

(5) (secondary school, 3-rd and 4-th grade, age 17-18):
Let ABCD be a tetrahedron and denote by K,L the midpoints of two of its opposite edges. Prove that every plane containing the points K,L divides the tetrahedron into two parts with equal volume.

(6) **(secondary school, 3-rd and 4-th grade, age 17-18):**
Find all real numbers k with the following property:
if x,y,z are the lengths of the sides of a triangle, then

$$x.x + y.y + z.z \leq k.(xy + yz + zx)$$

For pupils from the 4-th to the 6-th form of our basic schools (aged 10 - 12) there exists one more competition in problem solving - the so called PYTHAGORIAD. Its character is quite different from that of the olympiad. In each round pupils have to solve 10-15 traditional school-problems (often taken directly from the textbooks). They need not give reasons for their answers, just put them down as quickly as possible. (No choices are submitted.) The promptness of supplying the solutions is evaluated as well: the time limit is one hour and for t saved minutes the contestant obtains a bonus of [t+4] points (each correct answer is worth 1 point, a necessary condition for getting the bonus is obtaining at least 8 correct answers.) Thus, after solving eight or nine problems (and being sure of the correctness of the answers), it is a question of tactics whether to continue or to finish and get a higher bonus. Sometimes the instantaneous condition (quick and right decisions) is more relevant than precise (but often slow) logical thinking and trained problem-solving strategies. This is what makes this type of competition attractive to a wider field of children, not only to the gifted ones. And that exactly is the aim of the Pythagoriad.

Here is a sample set of problems for 5-th grade pupils (aged 11):

(1) In the division 24836/7 change the first number so that the result will diminish by 100.

(2) On a meadow there are 73 girls. The number of boys is six times greater. How many children are there playing on the meadow ?

(3) Calculate 5.5.5.5.5.2.2.2.2.2

(4) How many digits has the product 547 × 328 × 601 ?

(5) How many right angles are there in the diagram :

(6) Set the following numbers in decreasing order:
   1375/5 , 1656/6 , 1953/7

(7) 3/5 of an hour - how many minutes is that ?

(8) Replace the stars by digits so that the following holds:
   5*7 / 9 = 6*

(8) A group of children collected 260 mushrooms together. 1/4 of them were poisonous, 1/5 of them were inedible (but not poisonous), the rest were tasteful ones. How many mushrooms could the children eat ?

(10) Find the sum of all whole number solutions of the inequality:
   11 < x < 40

(11) Draw 6 different straight lines so that they intersect exactly at seven points. (The intersection points outside the paper are considered as well!)

(12) Find the area of a square with a side 567 cm long.

(13) Solve the equation: 2023 - x = 724 + 119

(14) 13 kg of beans cost 91 crowns. How many kilograms of beans can be bought for 113 crowns ?

(15) Complete the table so that the sums of numbers in all indicated directions will be 54 :

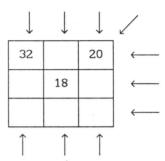

For those secondary school students who are interested more in deeper systematic investigation of advanced (even open) problems, a special competition called STUDENT'S RESEARCH ACTIVITY (SRA) is being organized. It is divided into several sections (according to the respective topics) one of which is devoted to mathematics and programming. A student (or a group of 2-3 students) interested in some kind of scientific work receives from a specialist (mathematician) a suitable problem. He tries to investigate it using mathematical literature and discussing the results regularly with his superviser. By the end of the school-year the student summarizes the results achieved in a written form and defends this paper against a committee of experts. These evaluate the professional quality of the paper as well as the form of elaboration and defence.

Here is a sample of problems which have been investigated by students within the SRA:

Problem 1. (Modification of the Old Chinese hypothesis)
Define a sequence $a(n)$ recursively:
$a(0)$ - arbitrary natural number,
$a(n+1) = a(n)/2$ if $a(n)$ is even
            $a(n)+b$ if $a(n)$ is odd   (b is an odd parameter)

Describe in detail the behaviour of this sequence in dependence on the parameters $a(0)$ and b.

Problem 2. (proposed by T.Hecht)

Let $p(x)$ be a polynomial with natural coefficients. Denote by sd [n] the sum of digits of the number n. Investigate the sequence sd [p(1)], sd [p(2)], sd [p(3)],... (in dependence on the polynomial $p(x)$).

Problem 3. In a plane with cartesian coordinates we introduce a non-euclidean metric - the so-called "postman metric":

For X [x(1),x(2)] and Y [y(1),y(2)] their "postman-distance" is defined as follows:

$$p(X,Y) = |y(1)-x(1)| + |y(2)-x(2)|$$

It is known that many geometrical objects and concepts can be easily defined using only the notion of distance (and a few other elementary set-theoretical notions): e.g. segment, circle, conic sections, equilateral triangle, perpendicularity, parallelism, etc. The question is :

What would these objects look like (and what properties would they possess) when replacing the "usual" metric by the "postman-metric" in their definitions ?

Now let us say a few words about the CORRESPONDENCE SEMINARS. They started some 12 years ago and soon turned out to be one of the most fruitful ways of discovering and fostering talents especially for those living outside the main centres. Children who show interest are regularly (approximately once a month) receiving by mail sets of 4-5 problems. They have to solve them within 3-4 weeks and send the written solutions to the organizers. These correct them and every participant is sent his solutions back with detailed comments from the makers. Attached (s)he also finds complete solutions of the problems and a ranking list of all participants.

The competition consists of 6-9 such sets during a school-year. Two or three times in a year the best 30-40 participants are invited to a one week meeting. On the programme of such a meeting there are various attractive mathematical activities but also sports, games and recreation. It is perhaps interesting to mention that, although the mathematical programme of these meetings is not obligatory, as a rule it is attended by almost all participants. The atmosphere, the climate of these meetings as well as friendships arising there among children with common interests, turned out to be a surprisingly efficient motivation for children to attack even hard problems and to study mathematical literature. At present we have about 15-20 working local correspondence seminars in Czechoslovakia which cover the age-groups 10-18. In contrast to MO and SRA they are not centrally organized. They arise spontaneously in those regions where a group of enthusiastic organizers comes together and some financial support from local authorities can be obtained.

It is not surprising that these seminars positively influence the results of their participants in MO and provide a sufficient number of well-qualified candidates for the special classes mentioned above. Perhaps a more interesting fact is that many successful participants of these seminars, when already studying at universities, show much interest in taking part in organizing these seminars. Thus, there is a nice continuity in this activity.

Here are a few problems from the correspondence seminar called PIKOMAT (created and for 5 years led by the first author). Its participants are 7-th and 8-th graders (aged 13-14) from basic schools in Bratislava and Western Slovakia (two districts of CSSR):

(1) In an empty room a great ball (with radius 1) is "running after" a smaller one wishing to crush it. What is the greatest possible radius of the chased ball if it wants to save itself in a corner of the room ?

(2) Let's call a natural number n "interesting", if 1 can be expressed as a sum of n distinct unit fractions. Thus, for example, 3 is interesting, because

$$1 = 1/2 + 1/3 + 1/6.$$

Find all interesting natural numbers.

Competitions in Czechoslovakia

(3) In the leftmost square of a board 1 × 1986 (a strip of 1986 squares) there are three buttons. Two players are playing the following game: they move alternately the buttons towards the rightmost square. In his turn a player is allowed to shift any (one) button rightwards to any other (even occupied) square. A player unable to move because all buttons are in the rightmost square is the loser. Find the winning strategy for one of the two players.

(4) An 30 × 40 × 50 rectangular cuboid consists of 60 small cubes, 10 × 10 × 10 each. How many of these cubes are intersected by the longest diagonal of the cuboid ?

The following problems are taken from a correspondence seminar for secondary school students (aged 15-18):

(5) For $A, B \subset R$ denote by $A+B$ the set $\{a+b\ ;\ a \in A,\ b \in B\}$. Prove that if $A+A = A$, then A contains a sequence $a(n)$ such that $a(n) \longrightarrow 0$. Further decide, whether from the fact that A contains both positive and negative numbers it can be deduced that A also contains 0.

(6) Let ABC be an equilateral triangle and P an arbitrary point in the plane lying not on the circumcircle of the given triangle. Prove that there exists a triangle with sides of length $|PA|, |PB|, |PC|$.

Besides these main MC some other local ones are being organized. These comprise competitions in various journals, in TV, on radio and competitions organized by particular schools only for their own pupils, etc. But, as already mentioned above, these don't have great importance.

Finally let us say a few words about the problems and perspectives of MC in Czechoslovakia.

The present tendency is not to increase the number of nation-wide centrally organized competitions. One of the reasons is that many teachers complain of the disturbance of lessons by a great number of various competitions (not only mathematical ones). (We don't subscribe to this opinion. According to our experience a competition should always be viewed as an efficient help to the teacher because its impact on the pupils is usually much stronger than that of an "undisturbed" school-lesson.) Our teachers are quite overworked and so they see in the competitions only an additional work. Some kind of popularization of MC among our maths teachers would be necessary.

Another problem is that of getting financial support for organizing MC. These can be sponsored only by local educational authorities or by our youth organization. But these usually are "poor" sponsors which, in addition, are not always aware of the necessity of various MC. Thus, the present tendency is to run small local MC, unpretentious as regards the finances, so that most of the organizational work can be done by a small group of enthusiastic volunteers. The effect of this kind of work is surprisingly strong. Close and friendly relationships which arise between the organizers and the participants of "their" competition not only strongly influence the pupils' views on mathematics, but sometimes also their plans for future studies.

Much more could be said about the Czechoslovak competitions and about MC in general, but the paper would become too extensive. We also didn't mention other out-of-class activities which contribute to PoM in Czechoslovakia (e.g. our mathematical clubs and summer camps of young mathematicians,...). We hope that the seminar will provide further opportunities to share these experiences with those interested. The authors would appreciate any exchange of information on this topic and are prepared to supplement this brief survey with additional information and experiences.

Notice:
The views presented in this paper are those of the authors and they should not be considered as an official report on the situation in PoM in Czechoslovakia.

Games and Mathematics

MIGUEL DE GUZMÁN

Universidad Complutense, Madrid.

"Can we analyse the relation between 'savoir faire' in puzzles and games and mathematical modes of thought? If we use such methods of popularization, how do we prevent mathematics from being associated with the solution of inconsequential problems?" (A.G.Howson, J.-P.Kahane, H.Pollak, <u>The Popularization of Mathematics</u>).

1. MATHEMATICS, ART AND GAMES

Mathematics is a many-sided human activity. It is, of course, a science; even more, it is the model and paradigm of all scientific activity. It is a powerful instrument for the exploration of the universe and for the appropriate use of the natural resources at our disposal. It is a model of thought which along the centuries has served as a privileged field for the study of the capacities of the human mind.

But mathematics has also been and continues to be an authentic art and game and this artistic and gamelike component is so consubstantial with the development of mathematics that every field of mathematical work that does not attain a certain level of aesthetic satisfaction remains unstable, reaching for a more polished expression that might convey a unitary, harmonious, pleasurable, amusing vision, in the same way as an unfinished symphony or poem stretches out in the mind of its author for the most beautiful possible form.

In what follows we shall briefly analyze the relationships of mathematics with games, leaving aside the study of the artistic components of mathematics, which has been carried out by many authors, among others Garrett Birkhoff, Helmut Hasse, Andreas Speiser, Hermann Weyl,...

From this analysis a germ of an answer to the second of the questions formulated by Howson, Kahane and Pollak will hopefully emerge. For this purpose one should try to carefully delimit the term "inconsequential problems". As we shall see, it is certain that many of the problems that arise from games are very far from being inconsequential. Many of the greatest mathematicians along the centuries have understood it this way and have devoted to them an intense effort, convinced as they were of the mathematical value they had.

2. THE NATURE OF GAMES

Games have been analyzed in depth by the sociologist Johann Huizinga in his work Homo ludens. He emphasizes the following features as characteristic of a game:

*The game is a free activity, free in the sense of the Greek paideia, i.e. an activity which is exercised for the sake of itself, not for the profit derived from it.

*It has a certain function in the human development. The human cub, like the animal, plays and prepares himself for competition and for life. The human adult plays and by so doing he feels a sense of liberation, evasion, relaxation.

*A game is not a joke. Games have to be played with a certain amount of earnestness. The worst game spoiler is the one who does not take it seriously.

*The game, like the work of art, produces pleasure through its contemplation and execution.

*It is separated from ordinary life in time and space.

*There are certain elements of tension in it, whose catharsis and liberation cause great pleasure.

*The game gives rise to very special bonds among practitioners, a sort of deep brotherhood.

*Through its rules, the game creates a new order, a new life, full of rhythm and harmony.

A perfunctory analysis of mathematical activity allows us to check that, in many of its forms, all these traits are present. Therefore, mathematics is also, by its very deep nature, a game, although this game involves other aspects, like the scientific, the instrumental, the philosophical ones, that together make of mathematics one of the fundamental pillars of our human culture.

3. THE PRACTICE OF MATHEMATICS AND GAMES

If games and mathematics have so many features in common regarding their ends and nature, it is no less true that they also share the same essential traits in what concerns their practice. This is particularly interesting when one is asking for the most adequate methods to transmit to a wide audience the profound interest and the enthusiasm that mathematics can generate and to convey a first familiarization with its usual ways and procedures.

Any game starts with the introduction of a set of rules, a number of objects or pieces, whose function in the game is defined by those rules, in exactly the same way as the objects of a mathematical theory are determined by implicit definition: "We are given three systems of objects. The objects of the first system we call points,..."

Whoever gets started in the practice of a game has to acquire a certain familiarization with its rules, relating the pieces with each other in the same way as the novice in mathematics compares and makes the first elements of a theory interact among themselves. These are the elementary exercises of a game or a mathematical theory.

The practitioner who advances in the mastery of the game is capable of acquiring a few simple practical techniques that, in circumstances which appear rather often, lead to a successful end. These are the basic lemmas and facts of the theory that usually are easily accessible in a first tackling of the easy problems of the field.

A deeper exploration of a game with a long history will give the practitioner a knowledge of the particular ways and procedures that the true masters of the game have left to posterity. These are the moves and strategies at a deeper and more complex level that have required a special insight since they are far from the initial elements of the game. This corresponds in mathematics to the phase in which the student tries to assimilate and make his own the great theorems and methods which have been created throughout the history of the subject. These are the thinking processes of the truly creative minds which are now at his disposal in order that he also can find light in the middle of confused and delicate situations.

Later on, in the more sophisticated games, where the stock of problems never becomes exhausted, the advanced player tries to solve in an original way situations of the game that have never before been explored. This corresponds to investigation of the open problems of a mathematical theory.

Finally there are a few who are capable of creating new games, rich in interesting ideas and situations that give rise to original strategies and innovative styles of playing. This is parallel to the creation of new mathematical theories, fertile in ideas and problems, possibly with applications to other open problems and to revealing more deeply some levels of reality that until now have remained in the shadows.

4. THE IMPACT OF GAMES ON MATHEMATICS

Very frequently in the history of mathematics an interesting question made in a gamelike manner or an ingenious observation about an apparently innocuous situation has given rise to new modes of thinking. This is the sort of spirit that makes science advance effectively, when one is able to look at the subject in an unconstrained and playful mood, away from the severe and earnest context in which official science is usually placed.

The beginnings of combinatorial analysis are situated in the Book of Changes (I Ching) with its distribution of different divinatory symbols and in the construction also in China of magic squares with mystic connotations.

The games with stones (psefoi) of the Pythagoreans gave rise to interesting theorems in the theory of numbers. Zeno's paradoxes should probably be read as a mockery against the prevailing ways of thinking among contemporary mathematicians. Euclid himself used a collection of fallacies in one of his lost books, Pseudaria, as a means to motivate his students in correct thinking processes. Archimedes, with his Problema bovinum and his Sand-reckoner faces strange situations in a gamelike manner in order to sharpen his mathematical instruments.

The list of mathematical objects that have come into existence motivated by the spirit of games would be without end. It is enough to quote some of the names of important mathematicians who can be thought of in this context: Fibonacci, Cardano, Fermat, Pascal, Leibniz, Euler, Daniel Bernoulli, Gauss, Hamilton, Hilbert, von Neumann,... A short, but very rich sketch of the evolution of mathematical recreations can be seen in the article by W.L.Schaaf on Number games and other mathematical recreations in the Encyclopaedia Britannica.

5. MATHEMATICS IN GAMES

The richness of mathematical themes in classical and modern games is impressive. The best way to perceive this is to look through the classical works of Lucas or Ball (Ball and Coxeter) and through the bibliographical compilations on games made by W.L. Schaaf and published by the National Council of Teachers of Mathematics.

Besides arithmetic, geometry and number theory as traditional sources of recreations one can also name topology, combinatorial geometry, graph theory, logic, probability theory,... In all these old and younger fields there are uncountable open problems of an amusing and attractive appearance that are possibly as easy to state and as difficult to solve as, for example, Fermat's last theorem. These await the creation of new thinking processes that can throw some light on their solution. About many of them one could not say if they should be classified as serious mathematics or else as idle oddities or puzzles. One can surely affirm that any game or puzzle with enough depth can have very intense repercussions on interesting aspects of mathematics. In the creation of puzzles or games man can display his imagination with complete freedom without being constrained by the conceptual or methodological bonds of a traditional theory.

Among the many examples of games that could be presented to show the similarity of the thinking processes needed to explore mathematics and games or puzzles, I will describe one that strikes me particularly for the different levels of treatment it permits, from the merely manipulative to the profoundly mathematical, and for the richness of heuristical methods that can be brought forward in its exploration.

One considers an infinite board as in the figure, where a distinguished horizontal line h is given. At the beginning of the game one places an arbitrary number of counters in some of the cells of the board, at most one in each cell below the line h. One then starts moving the counters according to the following rules: One counter can jump over another one located on the adjoining cell at its left-hand side to an empty cell. The counter which has been jumped over is taken away from the board. For example, the counter in

Games and Mathematics

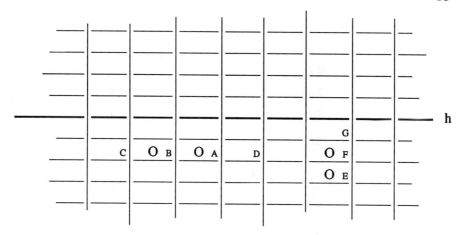

cell A of the figure can jump over B to C and the counter B disappears from the board. In the same way a counter can jump to the right-hand side (the counter on B can jump to D and then the counter on A disappears) and upwards (counter on E can jump over F to G and counter F is taken away). The game consists in trying to find an answer to the following question: What is the minimal number of counters one needs to place at the start under h and how should one place them to be able, moving them according to these rules, to place one counter on a cell in the fifth row above h?

6. THE SPIRIT OF GAMES IN MATHEMATICS

There is very much deep mathematics with the flavour of games. Among modern examples one can select a few in which this is very obvious. Some of them can be used in fact as a basis for amusing and entertaining games.

*Four colour theorem. Every planar map can be adequately coloured with just four colours.

*Ramsey's theorem (elementary version). Given six points on a circumference one joins each pair of them and paints the resulting segment blue or red. Then at the end there is always at least one triangle of such segments with the sides of the same colour.

*Sperner's lemma. One triangulates a triangle ABC, i.e. one partitions the triangle ABC into smaller triangles so that each two of them are disjoint or have only a side in common or only a vertex in common. The vertices of the triangulation are given the names A,B,C, under the only restriction that there should not be a vertex C on the side AB of the big triangle, no vertex A on BC and no vertex B on AC. Then, at the end, there is always at least one triangle of the triangulation whose vertices are A,B,C.

*Kakeya's problem. To find the infimum of the areas of all plane figures in which a needle of length one can manoeuvre in a continuous way so that at the end it occupies the same position in inverse orientation.

*The triangular billiards table. The ideal trajectories of a billiards ball in a rectangular table are either periodic or else they densely fill the table, i.e. they come arbitrarily near any given point. What can be said about the trajectories in an arbitrary triangular billiards table?

Many other problems could be quoted having the same ludic flavour: fixed point theorems, Helly's theorem, Hadwiger's conjecture, Borsuk's conjecture,...

7. MATHEMATICAL GAMES AS AN INSTRUMENT FOR THE TEACHING AND POPULARIZATION OF MATHEMATICS

Martin Gardner has assessed the situation quite rightly: "Surely the best way to wake up a student is to present him with an intriguing mathematical game, puzzle, magic trick, joke, paradox, limerick or any of a score of other things that dull teachers tend to avoid because they seem frivolous" (Mathematical Carnival, Preface).

The expert mathematician starts his approach to any question with the same spirit as a child starts playing with a new toy, open to the surprise, with deep curiosity before the mystery he hopes to illuminate, with the pleasant effort of the discovery. Why should we not use the same gamelike spirit in our pedagogical approach to mathematics? A well selected mathematical game can lead a student of any level to the best point of observation for each one of the subjects he has to face. The benefits of so doing are many: blockbusting, openness, motivation, interest, enthusiasm, amusement,...

On the other hand the similarity of structure of mathematics and games allows us to begin exercising in the games the same tools, the same thinking strategies that are useful in mathematical situations. Specifically the heuristic abilities in mathematics can be successfully initiated with the practice of many different games, as has been beautifully shown in the work of Averbach and Chein, Problem solving through mathematical games, with a very rich collection of games.

But above all, this gamelike approach to the more serious mathematical subjects can deeply benefit the student and positively influence his whole attitude towards diverse mathematical situations for the rest of his life, by showing him how to set himself in the right spirit in order to face mathematical problems.

From the point of view of the popularization of mathematics, the effectiveness of mathematical games is so obvious that I need not stress it. In the dedication of the recent masterwork by Berlekamp, Conway and Guy, Winning ways for your mathematical games, the authors write, with entire justification, "To Martin Gardner who has brought more mathematics to more millions than anyone else". Mathematics and games, as we have seen, are often indiscernible in their contents, but even much more in the common spirit with which they can be approached.

Mathematics is a great and sophisticated game that, besides, happens to be an intellectual work of art bearing at the same time an intense light to explore the universe and so having great practical repercussions. The attempts to popularize mathematics through its applications, its history, the biography of the most interesting mathematicians, through the relationships with philosophy or other aspects of the human mind can serve very well to let mathematics be known by many persons. But possibly no other method can convey what is the right spirit of doing mathematics better than a well chosen game.

Mathematics and the Media

MICHELE EMMER

Universita della Tuscia, Viterbo, Italia

1. INTRODUCTION

In recent years, there has been a significant increase in the amount of interest shown in scientific matters by the mass media, in particular by newspapers and television. The general public is showing considerable interest in scientific activities in the fields of physics, chemistry, biology and medicine. One only has to think of a recent example like nuclear fusion, and the enormous amount of interest that this subject generated in newspapers and on television all over the world. To what extent the information was scientifically correct is, of course, another matter.

In the fields of physics, chemistry, biology, medicine and natural sciences, we can say that there is constant coverage by the media of all the latest developments and the most recent researches. There is no doubt that one of the main reasons is that some of the results of scientific activity in these areas have immediate effects on the world community.

Another important factor to be taken into consideration is that these fields of scientific research require massive financial investment; so it is to be expected that the scientists themselves also wish to point out the necessity of developing one field rather than another.

So we reach the situation, as in the case of nuclear fusion, where the news of the experiments is announced to the press before the full scientific report has been published in the specialized journals.

Research work is not simply a *private* aspect of the researcher's life. Rather, it is something that comes under the heading of what can be called the *politics of research* involving not just different groups of researchers in a single country but research centres in all the major countries.

Another important aspect, in my opinion, is the attitude of the research workers towards the history of their particular scientific discipline. People working in the fields of physics, chemistry, biology and medicine have a very different attitude to those involved in mathematics. It is quite normal for a physicist involved in research to have an interest in the history of that particular sector.

In a recent article Patricia Clark Kenschaft, of the Montclair State College, has written [1]:

"Too often mathematicians are considered to be cold and insensitive people. This picture has a negative effect on our ability to acquire research funds and scholarships, to recruit teachers, and to convince primary school teachers to learn sufficient mathematics to prepare students for a technological world. We have to change our *image*."

As pointed out in a recent motion presented at an AMS congress, the work of a mathematician is largely the work of a single researcher, or at most of small groups, who generally speaking do not need large sums of money for their research. In other words, mathematicians do not need to *get into politics*, unlike other groups of scientists who need large financial resources. The result of this situation is that the world's scientific Museums have large departments devoted to the history and development of physics, chemistry, biology, medicine and natural sciences. The task of these museums is clearly made easier by the fact that they are able to display tangible *objects* representing the evolution of a particular discipline. Also, many of these museums have special areas where visitors can actually watch demonstrators reproducing classic experiments from the history of sciences.

Having said all this, what then is the situation as far as mathematics is concerned?

2. THE ROLE OF MATHEMATICS

First of all, I would like to make a few observations about mathematics and the attitude of the general public towards this subject. It is obvious that the only contact that many people ever have with mathematics is at school. Generally speaking (and unlike other disciplines), mathematics teaching does not deal with the history of the subject, nor does it make any attempt to link the history and development of mathematics to that of other scientific and humanistic disciplines, thus showing their mutual interaction.

Mathematics is presented as a ready-built construction made up of a sequence of facts which are not open to discussion (since they are mathematically demonstrable!). While in other subjects it is normal to know the names of the scientists who have achieved fame in a particular field, in mathematics most people leave school knowing almost nothing about important mathematicians and their works (with only a few occasional exceptions).

Everybody knows that there are many different fields of specialization in say physics or medicine, very few people realize that there is not just one subject *mathematics* but a whole series of different specializations that flow into the wide river of mathematics.

This non-historical presentation of mathematics in schools means that mathematicians are considered to be academics who are cut off from the rest of the scientific world, without any exchange of information or mutual interaction. Then again, people get the impression that mathematics is a sort of science in which everything is already defined, in which the only activity is the addition of a few more theorems to those that already exists.

For most people, mathematics means simply arithmetic and the Euclidean geometry learnt at school. After leaving school, these people not only ignore mathematics but also question the role of mathematicians in today's world. What do mathematicians do in their research work? Such a question would be ridiculous if asked about a physicist, a chemist or a doctor.

The result of maths-teaching in schools is that most people know very little about what mathematics is today, about the different topics within mathematics, and about what mathematicians do. It is true to say that there are specific problems involved in the popularization of mathematics, problems that are different to those posed by other scientific disciplines. In short, most people leave school with a marked aversion to mathematics as a whole, without really knowing what it is.

During recent years, the use of computers in schools has become widespread. So much so that if you ask someone what mathematicians do, most likely the answer will be "They work with computers".

To all this must be added the sad fact that many mathematicians have little interest in the popularization of their science, and even less for its history, which is considered to be something almost irrelevant, something to be dealt with by those people who are not **real** mathematicians but only *teachers*. This of course makes popularization even more difficult.

One aspect of the problem that should not be under-estimated in my opinion is that in many sectors of modern mathematics it is extremely difficult to give a *tangible* idea of the question involved. This is partly due to the fact that each sector has its own language and techniques that are only comprehensible to the mathematicians working in that sector (obviously, it is impossible to explain such concepts to someone who lacks even a basic knowledge of the subject); but, whereas in other scientific disciplines it is usually possible to explain the purpose of an experiment, in many mathematical fields even this is difficult to explain to non-specialists. In many cases, it is not possible to answer the question "What use is it?" because this is not really the question.

It is difficult for an observer to understand that mathematics is indeed a science that seeks the solutions to problems, but that there is not much point in trying to distinguish between *real* (or applied) problems and *internal* problems of the discipline itself. In many cases the distinction between pure and applied mathematics is very contrived. In fact, theories and results that seemed to be highly abstract and specialized have sometimes been applied to very tangible problems, drawn from the physical world. There are numerous examples of this transfer.

One begins to realize how many problems there are facing the popularization of mathematics.

3. MATHEMATICS AND THE VARIOUS MEDIA

In the popularization of any discipline, it is fairly obvious that the methods used will depend in part on the nature of the medium - newspapers, films, exhibitions, conferences, etc. Over the last 15 years I have had the chance to experience all these sectors and I would like to make a few comments about each one. However, first there are several comments of a general nature that apply to all media.

In keeping with what I said in the previous section, it is clear that, in order to popularize mathematics to a wide and varied audience, one has to assume that most of the people involved do not have even the most elementary knowledge of mathematics, nor of the language used to express its concepts. Therefore, when working with the media, it is essential to provide the basic information and bibliographic references, without which any attempt at divulgation risks becoming a foolhardy undertaking. In addition, as I have already mentioned, it is important to select the subject matter carefully. Personally I am convinced that it is almost impossible to deal with some of the aspects of contemporary mathematics.

Therefore, when undertaking a project of this type, over and above the specific features and language of the medium employed, one has to bear in mind the problems created by the inherent difficulty of the argument, as well as the ignorance of the general public.

A fairly new development, over the last few years, in the popularization of mathematics is the itinerant exhibition. There are many examples of permanent exhibitions devoted to mathematics (in the USA, France, UK). But, over the last few years, we have seen the appearance of mathematical exhibitions linked to recent theories and results [2], [3], [4]. Images play an essential role in this sector, especially those created by computers with sophisticated graphic capabilities. Then another interesting aspect is that these events have become not just scientific shows but *artistic* displays.

Here the idea of building a bridge between scientific and artistic images enables one to deal, in a comprehensible manner, with a history of mathematics parallel to the artistic and scientific events with which the audience is more familiar. It is clear that I am referring in particular to the Dutch graphic artist M.C. Escher, [5] but the question does not finish there, as can be seen in the exhibition *The Eye of Horus: a journey into mathematical imagination* held in various Italian cities in 1989. [6]

Another interesting feature is that visitors to these exhibitions are presented with problems which they are asked to resolve, sometimes with the help of live demonstrators.

However, one should avoid reducing the matter to a question of *"mathematical games"*, whether complicated or less so. It is important not to give the impression that mathematics is merely a game; but at the same time it is equally important not to frighten the spectator. The aim should be to provide ideas and stimuli for looking closer at the subject.

4. MATHEMATICS AND FILMS

Films dealing with mathematics can be divided into two categories: brief movies, without sound tracks, used to illustrate elementary or simple phenomena; and longer movies with sound tracks and music. I feel that brief films have been completely superseded by personal computers. Nowadays, with computer animation techniques, one can create in real time, and directly on the screen, animation effects that used to be included in films with greater difficulty. So, I do not think there is much point in continuing to make films of this type. In any case, I suspect that their impact was minimum, except when the movies were made directly with the students. [7]

With so many images constantly surrounding us today, one of the problems we are confronted with is that of finding the right images to visualize situations in a wide range of scientific fields. In mathematics, we are dealing with ideas that are often abstract and difficult to grasp; obviously, we are not always able to find images that effectively clarify the question. This leads, in my opinion, to the conclusion that it is useless to attempt to create a sort of movie library covering every topic of mathematics. The cinema medium, where images have precedence over words, is clearly not suitable for such a task.

In order to use cinema techniques for mathematical subjects, the two most important aspects (the scientific facts and the images used to illustrate them) must arise from the same source. One cannot hope to decide on the subject first, and then to search for the images with which to visualize it.

In my personal experience, the decision to link mathematical subjects to the visual arts (to architecture as well as to physics, chemistry and biology) seemed quite natural. One of the aims of artists is *to make visible the invisible*; why not use the images that artists have created, starting from a more or less scientific base with the addition of a personal element, to talk about mathematics, and at the same time to talk about art while dealing with mathematics?

My main idea with regard to producing math-movies is that of creating cultural documents seen through the eyes of a mathematician. In other words, as I have written elsewhere [8]: "The movies are attempts to produce works which are, at the same time, vehicles of information of a scientific and artistic nature on various mathematical subjects, and also to stimulate the observer towards further investigations of these same topics. The possibility offered by cinema techniques are fully exploited. My intention was to use the full language of images and sound. The problem, of course, is to maintain a balance between entertainment and informative popularization in such a way that one aspect does not dominate the other."

The movie should not be a lesson with pictures, but rather a new language that integrates the two ingredients. Compared with other media, the cinema has the great advantage of being able to provide a large quantity of information in a limited period of time.

Another major advantage is that the language of the cinema is universal. The language of images in movement is understood by people of all ages and all cultural backgrounds. This aspect has to be fully exploited in movies concerning mathematics, which have to catch the interest of an audience ranging from primary school children, to university students and the general public.

Apart from the question of adaptations into different languages, one has to aim for a world wide audience, just like a normal full-length movie. In other words, the spectators have to be able to find that point of interest that stimulates them to want to know more, to go deeper into the subject, to invent their own ways of pursuing the subject. The principal aim of a maths movie is to help the viewers understand more, and to help them look closer at the material, in the light of their own knowledge and experience.

My own series of movies, *Art and Mathematics*, was based on these principles. Bearing in mind that the cost of making a 30 minutes movie is quite high, if it is not possible to reach an international audience (that is, by means of a presentation that is not linked to the educational programmes and cultural background of a particular country), then it is not advisable to use such an expensive medium [9].

I should point out here that when I refer to movies, I mean movies made on film. Video cassettes and the reproduction of movies on video cassettes is another matter. These techniques lead to more efficient use of the material from the educational point of view, but much of the visual impact is lost. My overall view is that the ideal maths movie should provide a certain quantity of information and stimuli (mainly visual) in a limited period of time, relating to a problem or a mathematical theme that is of potential interest for anyone.

An important aspect is that the techniques used should be the best possible, consistent with the level of financing available. Nowadays viewers are used to watching highly sophisticated and eye-catching images; they would be disappointed if a maths movie were not of a similar quality, not necessary at the same level, but definitely not a penny-pinching type.

In short, to be effective the maths movie must be produced using specialized cinema technicians, with a high degree of precision, bearing in mind that the viewers are in some ways prejudiced against the subject matter in question. In this context, I think back to my first contacts, years ago, with the programming executives of the Italian Television Corporation (RAI) when I first proposed making a series of movies on *The Cultural World of Mathematics*!

Of course, one has to find the right balance between the correct (though necessary incomplete) mathematical content and the spectacular effect. This is why I chose to make the series movie *Art and Mathematics* using the experience of both mathematicians and artists who had dealt with the same mathematical themes, visualizing some of them in their own ways.

One final remark: one should not fall into the trap of thinking that by just watching maths movies one can learn the subjects dealt with. The movie works if it helps stimulate further interest, without boring the viewer.

5. MATHEMATICS AND NEWSPAPERS

In December 1987, a congress organized by the French Mathematical Society was held in Paris, on the theme: "Mathematics for the Future: what mathematicians for the year 2000?". Even by French standards, the press devoted an unusually large amount of space to the event. For instance, on the opening day of the congress, the daily newspaper *Libération* produced an EIGHT page special insert on mathematics. The front page of the paper carried a box giving high visibility and promotion to the insert. It is worthwhile reproducing the whole text of the box, under the title *Objectif Maths:* "Mathematics is not well-known. But it is becoming more and more important in informatics, aeronautics, in the economy and in medicine. A group of French mathematicians, tired of seeing their disciplines misunderstood and used only as a mean of selection, have decided to become *seducers* during the congress *Mathematics for the Future* in order to catch the attention of politicians and industry chiefs. Explore the planet Math in our supplement."

The eight pages of the insert were illustrated with photos which one either liked or disregarded completely. However, they represented an imaginative effort to visualize what was written in the caption of the first photo: "The search for the unknown." During the Paris congress, a round table was devoted to the relationship between *Mathematics and the Media*: newspapers, television and scientific magazines. The round-table was organized by the scientific journalist, S. Deligeorges, and amongst the participants, apart from several mathematicians concerned with popularization, was the woman journalist who had prepared the *Libération* supplement on mathematics with other colleagues.

She said clearly that when she and her colleagues (all whom had some type of scientific or mathematical background) had proposed the idea, they were thought to be out of their minds. The main reason for such a reaction was that there is no such thing as generally accepted mathematical knowledge, so that when you deal with mathematics in a newspaper you have to start from the beginning every time. Another objection was: What are you going to talk about? What illustrations can you use? And the crucial question: But will anybody be interested? A question of some importance for a daily newspaper.

I have given a detailed account of the case of this French newspaper because it seemed to be a unique example. For the last two years, I have worked as a freelance collaborator with an Italian daily newspaper which is distributed on a national scale [10]. It is the only paper that has an entire page every day given over to science. On average, I wrote an article every twenty days, with a half page of the newspaper at my disposal (more or less five or six standard A4 pages). So my comments on the subject come from first hand experience. However, I will not deal with the monthly magazine, at a high level like the *Scientific American*, of which many other people have direct experience.

One of the problems that arise when writing for a newspaper is that the subject has to be linked to some topical, day to day situation. This is not too difficult with other scientific disciplines, but with mathematics it can become very complicated. It is difficult to talk about a *discovery* in mathematics, and to show how it was resolved. Also, there is no way that complicated symbols can be used in a newspaper. So, there is always the risk of over simplifying everything and not communicating anything.

Articles about mathematics must necessarily be partly free from topical restrictions, even though they should have some reference to a recent event (a congress or conference, an important result, etc.). They also have to be long enough to give fairly detailed information, including the basic bibliography on the subject. In my experience, an article should be about five or six pages long. Longer articles can be split up into installments. The language should be accessible without over-simplification, one should not hesitate to use a few words that not all the readers will understand. The criteria are similar to those that apply to articles on literature and philosophy; it is better to be exact and difficult than approximate and over-simplified.

In short, one should avoid giving the impression that everything about the subject can be said in five pages! If the articles appear with a certain regularity, a useful relationship builds up with the readers, leading towards a degree of mathematical awareness, so lacking today. One should avoid being over precise, while still aiming for the right degree of exactness.

For example, in July and August 1989 I wrote four articles on the number π. My idea was to explain why you can very often see in newspapers reports of how it has been calculated to a new record number of digits. I wanted to explain several things: first that π is an irrational number, explaining what that means and giving a short account of the very long history of the problem; I quoted from all the original papers, giving all references and I was also able to write the final result giving technical details on the demonstration; the second idea was to connect this problem with the other question of the non-possibility of squaring the circle. In this case also I was able to enter into details; finally I explained why mathematicians are continuing to calculate more and more digits of π in order to test new supercomputers. In all the papers were 24 pages long: equivalent to the length of a paper in the Scientific American, but written in a newspaper.

6. CONCLUSIONS

I have tried to show the best uses of each medium for the popularization of mathematics, pointing out the common features of the various languages, and basing my comments on my personal experience over a period of 15 years. In conclusion, I should like to say that in the case of films and exhibitions it is the image that plays the essential role, awakening interest and stimulating the imagination. Also newspaper articles aim to achieve the same effect. We have to create an "*awareness*" of mathematics using every available medium. But we should not be afraid of not doing *Real Mathematics*, but only popularizing mathematics. For real mathematics we need other tools, although it is the computerized image that is becoming more and more important.

REFERENCES

[1] Kenschaft P C. (October 1987) Notices of the AMS, Vol. 34, No. 6, p.897.

[2] Peitgen H-O & Richter P H. (1986) *The Beauty of Fractals*, Springer-Verlag, Berlin.

Calvesi M & Emmer M. (1988) *La geometria dell'irregolare*, Catalogue of the exhibition, Rome, Istituto della Enciclopedia Italiana, Roma.

[3] Hoffman D. (1986) *Embedded Minimal Surfaces, Computer Graphics and Elliptic Functions*, in Dold A & Eckmand B (eds.), *Global Differential Geometry and Global Analysis*, Springer-Verlag, Berlin, p.204-215.

Emmer M. (1987) *Soap Bubbles in Art and Science: from the Past to the Future of Math Art*, Leonardo, Vol. 20, No. 4, p.327-334.

[4] Francis G. (1988) *A Topological Picturebook*, Springer-Verlag, Berlin.

[5] Coxeter H S M, Emmer M, Penrose R, Teuber M (eds.) ($1986^1, 1987^2, 1988^3$) *Escher M C: Art and Science*, North-Holland, Amsterdam. See in particular Emmer M, *Movies on Escher M C and their Mathematical Appeal*, p.249-262.

[6] Emmer M (ed.) (1989) *L'occhio di Horus: itinerari nell'immaginario matematica*, book-catalogue of the exhibition, Istituto della Enciclopedia Italiana, Roma.

[7] For a catalogue of math movies see: Schneider D I. (1980) *An Annotated Bibliography of Films & Videotapes for College Mathematics*, The MAA.

Bestgen B J & Reys R E. (1982) *Films in the Mathematics Classroom*, NCTM, Reston.

Singmaster D, *List of 16mm Films on Mathematical Subjects*, Polytechnic of the South Band, London, various editions.

[8] Emmer M. (1989) Art and Mathematics: A Series of Interdisciplinary Movies, ZDM 1, p.23-26.

[9] Emmer M. *Art and Mathematics: an Interdisciplinary Model for Math Education*, in Blum W, Biehler J, Huntley I, Kaiser-Messner G, Profke L (eds.), *Applications and Modelling in Learning and Teaching Mathematics*, E Hordwood Ltd., Chichester, pp.213-218.

[10] From 1987 I have written 30 long articles (5-6 pages) and 10 short ones (2-3) in the newspaper *L'Unita*.

Square One TV : A Venture in the Popularization of Mathematics

EDWARD ESTY AND JOEL SCHNEIDER

Content Director and Principal Mathematics Consultant, Children's Television Workshop, One Lincoln Plaza, New York, NY 10023, U.S.A.

Mathematics is decidedly not popular in the United States of America. Course enrollments decline markedly as students progress through secondary school. Many of our students achieve little more than low-level computational skills. Adults often exhibit a narrow view of mathematics tinged with disregard, even hostility. Even though many adults recognize the importance of mathematics in work and careers, they often disavow its personal relevance. In fact, our businesses and industries spend billions of dollars each year on remedial programs. If one accepts the premise that mathematics is essential to a well-functioning citizenry, then popularization of mathematics is not only valuable but necessary. The general failure of our principal effort in the popularization of mathematics - pre-college education - suggests that popularization be attempted in alternative settings.

1. THE CTW MODEL FOR TELEVISION PRODUCTION

The Children's Television Workshop (CTW) produces educational television programs for children. With Sesame Street, its first production, CTW initiated the use of mass-market, commercial broadcasting techniques and styles for educational purposes. The CTW model for television production brings together three distinct groups: production, research, and content. The production group consists of the full range of specialists in television production. The researchers are specialists in child development, psychology, or communications. The content group consists of experts in the subject area of the field of the series at hand. The three groups work in concert under the leadership of an executive producer whose experience is grounded in television production, rather than any of the research areas or a content area. CTW has refined this model over 20 years in the production of Sesame Street, The Electric Company (dealing with the reading of English), 3-2-1 Contact (concerning science), and now Square One TV.

2. SQUARE ONE TV

Square One TV is a television series about mathematics. Its main audience are children voluntarily watching at home. Non-commercial stations broadcast its half-hour programs five days each week, Monday through Friday, usually late in the afternoon, after school. Square One TV follows the magazine format, that is, each program comprises a number of independent segments ranging in length from 10 seconds to 10 minutes. Within this format, the approach is to parody television broadcasting practices and conventions. A segment may be a parody of any of the programs and devices typical of the commercial channels: situation comedy, detective drama, music video, game show, news programming, commercial interruption, self-promotion, and so on. We discuss the rationale for using this format and approach below.

The Square One TV library now includes 155 programs. For these programs we have produced segments of several types:

Studio Sketches	253
Mathnet Episodes	95
Animations	207
Non-Studio Pieces	55
Game Shows	113
Musical Pieces	45
Total	768

Studio sketches are short pieces featuring a seven-actor repertory company. The pieces parody well-known television productions and television genres.

Mathnet is a collection of five-part serials, an episode of which ends each program. Filmed on location in Los Angeles and New York, Mathnet parodies a popular detective drama of the 1950's. A variety of animation techniques appear in the library. Non-studio pieces include films and videos shot on locations outside of our studios. They typically take the form of a commercial interruption or an interview. The eight mathematically-based game shows are very much in the style of popular game shows on network television. Some musical pieces emulate the popular music-video format.

3. AUDIENCES

The intended audience for Square One TV are 8- to 12-year-old children. Secondary audiences include teachers and parents of children in the primary audience, and other viewers, both children and adults. Individual broadcasting stations independently schedule the program, typically in the late afternoon when children are likely to be at home. Nielsen ratings show that, on the average, about one million households view the program each broadcast day.

An interesting note from the Nielsen information is that Square One TV has a significant number of viewers younger than the 8- to 12-year-old target group. Square One TV is also finding an international audience through broadcast licenses in 22 countries other than the United States. The programs air in their original version, except in Australia where segments which significantly feature non-metric units of measurement are replaced.

Even though viewers are usually at home, some also watch the program in school. CTW permits the school community to record the programs on video tape for use for three years after their broadcast. While half-hour programs are not often appropriate for formal instructional use, the magazine structure of the programs facilitates using shorter pieces. However, there are significant impediments to any use at all in schools. The necessary hardware is often not readily available. Many teachers are not prepared, nor do they have the time, to integrate the material into their programs. To encourage teachers, CTW provides three guides which describe instructional uses for a small number of segments. Nonetheless, even though we have anecdotal evidence that some teachers are making use of Square One TV, there is no indication of substantial use among the nation's 1.2 million elementary school teachers.

4. GOALS FOR THE SERIES

Square One TV has three goals. Its first and primary goal is to support and stimulate interest in mathematics among its target audience. Interest in mathematics appears to begin to decline in the late elementary grades; hence our choice of audience. CTW's experience with earlier productions indicates the potential effectiveness of television in affecting attitudes. In our approach to Goal I, we try to show mathematics as a powerful and widely applicable and useful tool; to present some of the beauty of mathematics; and to convey the message that mathematics can be understood and used by non-specialists, even those in our audience. In addition, we try to help the viewer recognize mathematics in out-of-school contexts.

Goal II is to model good problem-solving behavior. Characters in the series' segments encounter mathematical problems and deal with them willingly and successfully. Their actions illustrate aspects of problem formulation, problem treatment, and problem follow-up, while they apply a variety of problem-solving heuristics.

Goal III is to present a broad view of mathematics. School mathematics in our country is concentrated on computational arithmetic. Without the constraints of a standard curriculum or testing program, we are free to present any mathematics which we can render interesting and accessible to our viewers. Thus, we undertake elements of geometry, probability, statistics, combinatorics, and functions and relations. Standard topics such as properties of numbers and counting, arithmetic, and measurement also appear in problem-solving contexts. A detailed statement of the goals, together with an analysis of how the segments from all 155 programs relate to those goals, is available from the authors.

5. RESEARCH ACTIVITIES

We maintain an extensive program of formative research as an integral part of the production process. We show new segments to groups of target-age children to assess the material's appeal and comprehensibility.

In addition, we have recently completed a major summative study of programs from the first two seasons of Square One TV. Its purpose was to examine in great detail changes that might occur in children's attitudes toward mathematics (Goal I) and in their inclination to use problem-solving techniques (Goal II) as a result of sustained viewing of the series. The study was organized in a pre/post experimental/control design, with a treatment consisting of watching 30 half-hour programs from the series. At both pretest and posttest, subjects were individually interviewed in two 55-minute sessions, using a set of nonroutine mathematical problems. Their performance in relation to Goal II was measured by two scores, one involving the number and variety of problem-solving actions and heuristics used, and the other the mathematical completeness and sophistication of their solutions. Gains in both scores were significantly greater for the experimental group than the control group. Neither score, however, interacted significantly with gender or socioeconomic status (SES), indicating that boys and girls of varying SES were affected similarly by watching the series. Also, neither score was significantly correlated with scores on a standardized test of mathematical achievement that had been administered prior to the beginning of the experiment.

A summary of the problem-solving part of the study is available from the second author. A summary of the part of the study related to attitudes (Goal I) will be available in April 1990.

6. ISSUES RELATED TO POPULARIZATION

There are two arenas in which Square One TV can be considered relative to issues of popularization: external to CTW and internal to CTW.

Our main purpose is to convey to our audiences some notions about mathematics that are embodied in our official goals. As it happens, these ideas (e.g., mathematics as something that can be engaged in by non-specialists; mathematical problem-solving as something that is worthy of post-solution reflection; mathematical content as something that includes much more than the arithmetic that dominates school mathematics) are not widely shared by any of our viewing audiences - primary or secondary. Having decided to attempt to convey such a view of mathematics, the next steps included choices of specific content, situations in which to embed the mathematical ideas, level of mathematical sophistication, frequency of repetition and reinforcement, degree of complexity in presentation of concepts, among others. This decision-making process is heavily dependent, in turn, on popularization in the smaller arena - that consisting of our own colleagues.

Children's Television Workshop, as a company, is primarily engaged in television production. It has been producing television programs for children for more than 20 years; its top management is expert in, and almost exclusively concerned with, issues of television production. One principle of the CTW production model is to employ specialists in the many and diverse fields that contribute to a television show and have them cooperate under the leadership of an experienced executive producer, rather than to put the production under the direction of content experts.

At the peak of our work cycle about 55 people are involved in the various activities of production, research, and content. Several hundred more come in on an occasional basis in the execution of musical pieces, animations, and film work. These include members of our 18-member advisory board and other mathematical consultants on whom we call for occasional advice and assistance. Among this large group is a four-person Content Department. While we have substantial day-to-day input into the show, and responsibility for the mathematics, nonetheless it is humbling to keep in mind the fact that of all the many people who are directly involved with Square One TV, only the four in the Content Department come with a background in mathematics and mathematics education. In the main, the production staff and researchers, and so on, while eminently talented in their own fields, are outside of mathematics. Some have a casual interest in the field, but most fall along a continuum from neutrality through aversion and beyond. That is, even though skewed in terms of educational level and income, insofar as their feelings toward, and knowledge of, mathematics are concerned, they are much like any other random selection of U.S. adults. Of course, if the population at large were otherwise, there would be little reason for a project of the nature of Square One TV. Thus our first efforts in popularization are directed toward a small, clearly defined audience - our colleagues in television production.

To be successful, we must attract our principal viewing audience: children viewing at home. This is a largely discretionary audience. Many potential viewers have competing interests that might engage them as alternatives to watching Square One TV. In particular, there are other television shows that compete directly with Square One TV. Our children typically decide for themselves which programs to watch; a flip of the channel selector will expunge our show from the screen, no matter how mathematically worthwhile it is. To have any impact at all, then, we need first to attract an audience in a highly competitive environment. Hence the first criterion by which we judge any segment of our show is its appeal to our primary audience. Since our colleagues on the production side by and large do not find mathematics naturally appealing or attractive, they may find it difficult to perceive the mathematics itself as a source of appeal for the audience.

While this is frequently frustrating for us in the Content Department, there is a bright side to this picture from the point of view of effective popularization. To the extent that the staff of Square One TV is a microcosm of our external audiences, the Content Department's natural interest in and enthusiasm for mathematics is, in fact, unnatural. The fact that the Content Department must constantly promote, justify, and defend the mathematics which does appear on the show reminds us that we must be just as diligent with our viewing audience. To put it another way, one can imagine a popularization project in which all the staff were enthusiastic about, and well versed in, mathematics. Quite possibly the product could depend on the mathematics to be its primary attraction, and, as a result, fail to attract a broad audience.

The problem of producing an attractive, appealing series is exacerbated by the fact that our audience is greatly varied along several dimensions: age, location, experience, social sophistication, and mathematical sophistication, among others. This in part drives our choice of the magazine format and humorous parody as an approach. Our research shows that our audience is very knowledgeable about the conventions of the television. They watch large amounts of television programming. And they are attracted to broad humor and parody. Through the magazine format, we can aim to satisfy a diverse audience by producing individual shows whose segments vary in style, format, and tone, as well as in mathematical content and sophistication.

Other tensions in designing the series have to do with the mathematics itself. We regularly confront decisions about the amount and depth of mathematics in a segment or a show. One of the dangers of high mathematical density is in slowing the pace of the show to a crawl, especially in depicting arithmetic calculation. However, we do not necessarily avoid segments of high density. In fact, we have many pieces which are short and concentrated, as well as longer, more diffuse pieces, in which both the mathematics and the problem solving ebb and flow. In general, the more we work to develop plot and characterization, the more diffuse or uneven is the mathematics. However, since the audience prefers richer plot and narrative structure, we need these elements for their appeal.

Consider a specific example. One constantly hears of the importance of conveying the idea that mathematics is a useful and powerful tool, applicable to a variety of situations in the "real world". However, exactly how to do this remains a major problem, unresolved as far as we are concerned. In our first season we attempted a number of 3- to 4-minute segments that showed real people (that is, non-actors) engaged in occupations in which mathematics was used with varying degrees of explicitness. These included a piece on a tugboat captain who uses mathematics in piloting her boat around New York harbor; a professional basketball player concerned with bouncing angles; and various people in the wood-products industry. The mathematics in these segments was of a fairly high density, but they were not successful on any grounds: formative research showed that viewers did not like the pieces, content department opinion judged them to be ineffective, and the expert judgment of the producers rated them below standard for the series. We have tried other approaches, including a much lower-density exhortation to learn mathematics because of its usefulness in some specific occupation. We are not totally satisfied with this approach either, and we continue to look for effective ways to convey the ideas.

7. NEXT STEPS

As of January 1990, Square One TV has completed three rounds of production, yielding 155 half-hour programs. We are now beginning a fourth season in which we will produce 40 more programs for the library. In addition, we will explore an alternative format designed to promote family viewing. We expect to apply the lessons of the past several years to improve our technique and effectiveness. The challenge continues to be to attract and to hold our audience while conveying valuable messages that will expand and enrich their conceptions of mathematics.

Frogs and Candles - Tales from a Mathematics Workshop

GILLIAN HATCH[1] AND CHRISTINE SHIU[2]

[1]Manchester Polytechnic, All Saints, Manchester M15 6BH, U.K.
[2]Open University, Walton Hall, Milton Keynes MK7 6AA, U.K.

A Mathematics Workshop is an event organised for a group of voluntary participants - usually children - to meet and engage collectively in a variety of mathematical activities. In this paper we describe the development of mathematics workshops, mainly in the United Kingdom, by members of the Association of Teachers of Mathematics (ATM).

Workshops can take a variety of forms, perhaps the simplest of which is one mounted on a Saturday morning, when typically the organisers are members of a local branch of the ATM. Each of the contributing teachers will arrive with a couple of their favourite activities which they will run all morning with different groups of children.

Sometimes families arrive as groups, so that a wide range of ages of children work together with parents joining in as well. For example, at a recent workshop a set of mathematical games, available at several levels of difficulty, was set out on a table. It was observed that the older children were initially happy to play a simple game which everyone could cope with and then to help the smaller ones to play a harder version. At no point did they appear even to wish to take advantage of the fact that they were more competent at what was involved.

Other groups of children come with a teacher and the willingness on both sides to expend "free" time on the workshop indicates the value and appeal of mathematics presented this way.

Frogs and Candles

For more intrepid and devoted organisers, the residential workshop offers an extended chance to work in this creative way in a novel environment. This can allow not only the opening up of new mathematical avenues but, when teachers take their own pupils, the development of stronger pupil-pupil and teacher-pupil relationships.

The FROGS and CANDLES of the title are two activities which have proved perennially popular with children (and adults) in many workshop settings.

FROGS is the problem of two sets of frogs meeting as they use a line of lily pads as stepping stones to cross a pond. Each lily pad can support only one frog and the frogs pause when they see there is only one vacant pad separating the two parties. The aim is to interchange the two parties using the minimum number of permitted moves which are

(i) a glide (one frog moves to the adjacent empty pad), and

(ii) a leap (one frog leaps over another onto an empty pad);

and hence to discover the number of moves required by different sized parties. The activity is often initiated by two teams of children (boy frogs and girl frogs) enacting specific cases.

Whilst FROGS can be worked on anytime and anywhere, CANDLES is an out-of-door activity which needs the cover of darkness to achieve its effect, and is thus more used on residential workshops. Each participant carries a lighted candle, safely anchored in a glass jar, and moves to a position which satisfies the conditions given by the group leader. Several variations of increasing complexity are tried, and when the group leader sees that everyone is appropriately placed the participants place their candles on the ground and repair to an upstairs room from which to view the locus that has been mapped out.

1. WORKSHOPS AND THE POPULARISATION OF MATHEMATICS

As indicated above, mathematics workshops can take many forms but we believe that there are a number of characteristics which all successful workshops have in common. These include choice, enjoyment, activity, sharing and challenge.

Choice First of all the participants choose to come to the workshop, and, having arrived, find there is a variety of activities to choose from and a choice of leaders to work with.

Enjoyment A major aim of all workshops is for the participants to experience enjoyment in engaging in mathematical thinking. If the choice to participate is to be made there needs to be a promise of enjoyment, and to keep participants engaged, the promise needs to be fulfilled.

Activity Another emphasis is active engagement with the mathematics - workshops are based on the belief that mathematics is not a spectator sport. The starting activities are often very literally active starting from a "people maths" game or puzzle. There is usually plenty of physical apparatus around, but starters can be "in-the-head" problems.

Sharing Working in collaboration with others enables participants to contribute to the solution of problems which they could not tackle alone, and to share delight in group success.

Challenge The activities though enjoyable are not the end in themselves. There is typically a period of reflective follow-up to activities within the workshop itself, and usually further questions arise of the "I wonder what would happen if ...?" variety, which participants carry away and work on, sometimes long after the original event.

These attributes all contribute to the participants' developing image of mathematics. This image is "popular" both in the sense of being a shared image, evolving from the pooled perceptions of a group working together, and in the sense of participants being free to choose to engage in something enjoyable. They promote an image of mathematics as a dynamic and challenging field of activity rather than a static body of knowledge.

2. THE ATM CHILDREN'S WORKSHOP GROUP

In the late 1970s a number of members of ATM were experimenting separately with residential mathematics workshops for children. In 1981 a co-operative working group of members, interested in pursuing this idea further, was formed and met several times with financial support from ATM.

In July 1981 the group arranged and ran a five day residential workshop for children from two schools. The venue became popular and was the setting for several more similar, though often shorter events. These led to informal publications of children's work including the original "Maths at Simonsbath".

As the group collaborated with each other on developing their own expertise in setting up and running workshops, individual members were invited to give talks about their experiences to a variety of interested groups, often teacher groups who wished to run workshops themselves. It became evident that such teachers also welcomed practical support and so the group began to develop the notion of a workshop box. Two of these were set up, one in the north of England, and one in the south. They contained a variety of suggestions for workshop activities and the practical materials needed to support these. Teachers were able to borrow a box and use it as the basis of their own workshops, and were encouraged to add their own ideas to the contents.

By June 1982 the Children's Workshop Group offered interested members of ATM the following services, listed in their newsletter of that date.

"We can

1) keep you up to date with a regular newsletter with
   - information about workshops that are coming up
   - suggestions of how to set up your own workshop.

2) Offer you opportunities to
   - visit a workshop in action
   - bring some children to a workshop
   - set up your own workshop with help from experienced people.

3) We are putting together A BOX of useful materials which will be available for you to borrow.

4) Two pamphlets have been produced, written by children, based on their experiences at Simonsbath.
   - Maths at Simonsbath
   - Mathematics at large."

The development of the boxes eventually culminated in the publication by ATM, of a pack entitled "Away with Maths" which presents an extensive resource for running workshops.

3. EVALUATION

The ATM children's workshop group did not attempt a formal evaluation of their work. Instead they issued invitations to other teachers to observe and/or to participate, and hence discover the value of workshops for themselves. This theme will be taken further in the section entitled *Workshop* as *exemplar*. Here we simply remark that the group's success in spreading the workshop idea is some measure of value in itself.

Further evaluative evidence is manifest in comments made by participants. The children tended to remark on specific activities, (and it may be instructive to compare their descriptions of FROGS and CANDLES with ours), but often their written comments implied response to one of the five attributes of successful workshops which we have identified. For example:

"The thing I enjoyed most on this weekend was ... when we were allowed to go into one of the classrooms and make or do anything we wanted. ... Clare, Raphaella and I made hexagons with hexagonal beer mats. When we made a few we stuck them together. Then we tried to make stars out of matchsticks ..." (Choice, activity.)

"The FROGS is a puzzle where you have a certain amount of space and in each one of these you have to put a counter except for the central space. The frog (counter) is allowed to jump over another frog (counter) or move into a space next to it. The objective of the game is to get all the counters on the opposite side to what they started on. We got the answer correct and had a lot of fun in getting the answer."

"This was just one of the experiments which we had to work out. There were many other enjoyable puzzles which we had to work out. The weekend was very enjoyable and good fun."

"I think my favourite thing of all was with the candles. Everyone of us took a candle in a jar. ... Next we got a piece of paper and took it and the candle outside to where there were two lighted flares. Then we got two edges of the paper and lined them up with the two flares and placed our candle down at that spot. Finally we went back inside and looked out of one of the upstairs windows. We found out that all the candles had formed a large circle. It was a terrific sight." (Enjoyment.)

"The person in front had to put their right hand between their legs and join with the person behinds left hand and so on down the line. When that was done the person at the back had to go through everybodys legs and not break hands. Why don't you try it with a friend?"

> "I liked the tick-tock game because it was so stupid and everyone got muddled up with the tock going one way and a tick going the other and "a what!" somewhere in the middle." (Activity.)

> "One of my favourite problems was to do with cubes here with your group you had to imagine a cube, a white cube, and we had to count the number of sides on the cube the number of corners on the cube. We then imagined a line joining all the sides together through the middle of the sides and no more than one line on the side of the cube."

> "My favourite things on the maths weekend were games where we all got into groups." (Sharing.)

> "The four cubes puzzle is fairly complicated. I enjoyed this most of all because it keeps you buissy (sic) and makes you concentrate - hard." (Challenge.)

An older pupil, a lower sixth-former quoted in "Away with Maths", welcomed the opportunity to work at greater length and depth on some aspects of mathematics:

> "The work has given us all a chance to go much deeper into complicated problems, formulae and mathematics in general. We have been able to forget other subjects for a time and concentrate all our efforts on the maths. We have covered a large and varied amount of work in a very short space of time."

Group leaders as well as participants recorded reflections on their experiences. For example a student teacher said:

> "The informal, relaxed atmosphere of a maths weekend is something very special, but we have had the same "buzz" in lessons since then. ... A good and valuable time was had by all. Perhaps a weekend like this should be part of everyone's maths experience?"

A teacher who had reservations about taking part in any form of residential visit with children had her views modified when she took part in a well-organised residential Mathematics Workshop:

> "The first time I went away with a group of children to spend a weekend at a residential centre was in my first year of teaching. ... I had no previous experience of the sheer energy of children of this age group, and after an exhausting - if enjoyable time in parts - vowed never to do it again. Little thought had been given to the structure of that weekend, and I had no real place.
>
> I was 'captured' however after agreeing to spend a weekend away with a third-year tutor group. ... Much thought had been given to walks and activities and I had a place ... I was to do the evening sessions. A circle of chairs was placed for anyone who cared to join in. They were expectant and there was sufficient structure to demand involvement. ... I was aware that something different was happening here ... and the possibilities for maths times ..."

A number of teachers who pooled their reasons for running residential workshops, came up with the following list:

- workshops take maths out of the school environment and give children and staff a chance to have a real mathematical experience in a relaxed informal atmosphere.
- going away is exciting.
- children have the opportunity to pace and organise their own work.
- there are no bells to interrupt. They have time to explore their ideas to a greater depth.
- workshops motivate teachers as well as children. This may well be transferred back to the classroom.

- "it gives me a chance to do some exciting maths that is not on my syllabus."
- "it gives me a chance to do some of my school syllabus in an exciting way."
- teachers have the freedom to experiment and exchange ideas.
- the relationship between children and their teacher is altered.
- there are more adults.
- it gives children confidence to have a go.

4. WORKSHOP AS EXEMPLAR

The ATM Workshop Group concentrated on working with other teacher groups who contacted them. Other ATM members have used workshops to share their ideas with specifically targeted groups of adults.

Over a period of many years Manchester Polytechnic has run an inservice course for teachers of "middle-years" pupils (age range 9-13). This course is unusual in that its second half consists of a sequence of workshops which the teachers resource and run and to which they bring a small group of their own pupils. These run partly inside school hours and partly outside. From the children's point of view they appear to represent an exciting chance to come to a large "senior" institution and do some interesting mathematics. From the teachers' point of view they offer a chance to see the similarities and differences exhibited within a group of children of varying ages and from different schools. A vivid memory is the picture of a group of "Junior 4" children (aged 10-11) directing everyone else in an immense game of FROGS in order to verify their solution. Their competence and confidence were impressive and gave many of the secondary teachers an entirely fresh view of what young children can achieve mathematically.

Workshops also form a powerful means of changing parental attitudes. Some element of deception is perhaps needed the first time this is used, as only the confident will choose to attend a workshop if it is advertised as such. However experience has shown that those who find themselves at a "talk" which proves to be a workshop rarely, if ever, walk out again in protest, but remain to join in with good humour. Indeed at a recent workshop of this kind, when alas no camera was available, the sight of about a hundred parents deeply involved in games, puzzles and investigations remains vividly recorded in the mind's eye. There was a threat from some of the fathers to instigate a betting system for some of the more complicated games - perhaps we could have made a profit for school funds! However what was most memorable was that when the time came to give a short talk about the educational ideas behind the evening's activities, it was almost impossible to separate this large group of adults from the mathematics, indeed some of them continued to work on surreptitiously like naughty children.

We have asserted that describing workshops gives only the merest flavour of what they are like, and that it is better to see one for oneself, better still to participate in one. ATM therefore determined to put on a children's workshop at ICME6 in Budapest in 1988. Those who volunteered to run this were experienced and enthusiastic about running workshops in the UK, but were aware that they could meet barriers of language and cultural expectation in transplanting the form to Hungarian soil.

So a year before ICME6 a small advance party (Jan, David and Gillian) went to Budapest to establish contacts and to work with children in a school noted for its teaching of English. They discover that FROGS retained its almost universal appeal, and with just a little help at the start to set up the problem, they could work almost without need of words with a large group of children who knew only a little English. (They did however find it expedient to learn the Hungarian for "frog" and for "backwards"!) Indeed photographs taken of this large group of 80 or so children at work look almost disappointingly like any other workshop run anywhere, giving both a sense of anti-climax and of triumph.

Thus ATM ran its workshop at ICME6 with a core of some of the children they had met on the preliminary visit (who incidentally gave up four days of their summer holiday to participate). These children were joined by a growing band of children of conference members and together spent long tracts of time working on mathematics of all kinds, undeterred by the variety of languages being used. Typically to be found in the workshop, was a group of children of six or more nationalities, ranging in age from 5 to 14, working on activities. This usually also drew in the occasional parent, Hungarian teachers and other conference members who had come to see what was going on. "People games" (i.e. those games and puzzles which can be presented using people as counters) were particularly successful. The success of the workshop for the children is perhaps summed up by the very able boy, not a native speaker of English, who said at the end of the last day:

"Thank you for all the problems - at school they only give us calculations to do!"

There was only one aspect of ATM's ICME6 experience which was disappointing. Although we had been given plenty of space for both noisy and quiet activities, and for displaying our materials, we were neither in the main conference building nor the exhibition building, so many conferees passed us by. It was therefore decided that if the conference would not come to the workshop the workshop should go to the conference. So, ably led by "Jan the Magnificant" the children, and not a few adults, marched on the foyer of the main conference building and worked on some "people maths" activities there. This provoked more camera clicking than any other conference event we witnessed. The message went round the world. Doing mathematics is an international form of communication!

5. SUMMARY

In this paper we have tried to describe the nature of mathematics workshops and indicate their potential effect on the image of mathematics held by all those who become involved in them. We reiterate our belief that the best way to learn about workshops is to take part in one. In an earlier section we listed the characteristics of a successful workshop, but these were somewhat abstract so we conclude by offering a more detailed list of typical elements of mathematics workshops.

(i)   They are fun.

(ii) They can involve children of all ages and abilities.

(iii) They allow children and adults to work together on related activities.

(iv) They can involve activities which make sense to the smallest child but which can challenge the ablest adult.

(v)   They are active and practical.

(vi) They offer the opportunity to work on mathematics for an extended length of time.

(vii) Existing, as they do, outside the confines of the normal school day they convey the unspoken message that doing mathematics is a worthwhile way to spend time.

REFERENCE
The ATM pack "Away with Maths" is available from the Association of Teachers of Mathematics, 7 Shaftesbury Street, DERBY DE3 8YB, UK.

Mathematics in Prime-Time Television: The Story of Fun and Games

CELIA HOYLES

Department of Mathematics, Statistics and Computing, Institute of Education, University of London, 20 Bedford Way, London WC1H 0Al, U.K.

1. FROM IDEA TO TRANSMISSION

Yorkshire Television, one of the larger independent television companies in the U.K., is well known for programmes popularising science. In October 1986, the science department came up with the idea to try to popularise mathematics in a similar manner. As Duncan Dallas, the head of the science department, wrote "As a kamikaze notion it could hardly be better ... (Mathematics) is the least popular or accessible of the sciences. It does not sell magazines in the same way as computing, it is not the subject of dinner party conversations as is ecology, nor is it a trendy part of our life style like technology. Indeed, it is universally acceptable to trumpet our ignorance whenever the subject is mentioned. Clearly mathematics is important enough to command our attention but on a list of programme ideas rated by popularity it will probably become bottom" (Dallas 1988).

The crucial question was to find a format, a way into the mathematical perspective that would be entertaining and lively. After considerable discussion, Yorkshire Television came up with the idea to base the programme around puzzles and games. The rationale was to capitalize on people's interest in puzzles, an interest which goes back for many years, and to use these puzzles as a vehicle to think about the embedded mathematical ideas - after all recreational mathematics has been the source of a great deal of mainstream mathematics. The programme was called Fun and Games and it was to be transmitted at prime-time, seven o'clock in the evening.

The next step was for Yorkshire Television to decide on the method of presentation of the games and puzzles - another crucial question. The programme was to be set in the studio with a participating audience. The games and puzzles were to be active and 'larger than life' so that viewers also could be involved. The puzzlers would be 'ordinary' - the show was not to be a competition to discover 'egg heads'. It was also decided that there would be two presenters, a host and a mathematical expert, a 'professor', to explain the mathematics. The host was to accompany puzzlers through the games, set them at ease, explain the rules, be witty and friendly and liaise between audience, puzzlers and the 'professor'. The 'professor' was to be knowledgeable about the mathematics, demonstrate strategies, and exude enthusiasm about abstract principles, intriguing mathematical ideas and their application! Eventually I was asked to be the 'professor' and to present the programme with Johnny Ball, a TV comedian, well-known for his television programmes concerned with children's mathematics. When I was first approached I was most excited by the prospect of doing the programmes, indeed, the audacity of the idea! I have long been aware of the pervasive negative image of mathematics. Indeed, in my inaugural lecture I stated that "The majority of pupils are anxious about mathematics, alienated from it, or simply bored by it. They rarely become involved in the subject, treating it as something to be got through rather than understood. Even 'successful' pupils have been found to harbour incorrect ideas at the most fundamental level. Mathematics urgently needs, therefore, to be made more accessible and meaningful for pupils" (Hoyles 1985). Schooling seems to turn people off mathematics and the majority of the population do not see any way that the subject can be interesting and intriguing, let alone fun. Their view of mathematics is of rows of sums or meaningless rule-following. Mathematics does not relate to people's intuitions nor does it appear to have any coherent structure in itself. Formal representations are imposed 'from above' in school and these formal representations thus have little meaning and less interest.

I therefore was delighted to join the programmes and to try a new approach to the problem of popularisation of mathematics. I suppose I had a vision that people would, around dinner tables or over a drink, try one of our puzzles themselves, think about the mathematics and, perhaps more importantly, talk about it. Thus another view of the nature of mathematics could be glimpsed by more than the present tiny élite minority. I also acknowledged the need for a female in the role of mathematician given the enormous literature on gender stereotyping in mathematics (see for example Fennema 1985) and the importance of presenting a positive role model for women. The format for the first series was as originally conceived and transmitted in July 1987. Much to everybody's surprise, and, indeed, delight, the programmes were a great success and the series achieved audiences of between eight and nine million during high summer. In fact the programme reached the ITV's Top Ten and was widely heralded as "streaking ahead of its great Wednesday night rival". (Daily Mirror August 1987).

I think it is true to say that Fun and Games is the first television programme to try to use mathematics as a focus in a prime-time slot. Sometimes the mathematics is simple, sometimes quite hard but we always firstly strive to highlight the mathematical principles employed in the puzzles and secondly have fun! We want to challenge people's long-held conceptions of what mathematics is all about by provoking involvement in the subject in an enjoyable setting and bringing in the surprise factor - the 'Aha!' when people suddenly see a new way through or when they look at a problem differently - that is, mathematically! So to recapitulate, what were our aims? To show that mathematics is not just 'boring old numbers'; to show how mathematics can be the key to solving puzzles; to show that mathematics can be fun, intriguing and perhaps exciting; and, finally, to infiltrate mathematics into the general popular culture.

There was another series of Fun and Games in the summer of 1988 where I was joined by a new co-presenter, Dr Rob Buckman and a third series in the spring of 1989. The audience figures are holding up and we now have loyal followers!

2. THE STRUCTURE OF FUN AND GAMES

Some things about the programme have not changed over the three years since its first transmission. There is still no competition and no extrinsic incentives such as rewards or prizes. There are still two presenters, myself and Rob Buckman. Rob's role is crucial in providing a clear exposition of the puzzles and in keeping the atmosphere of the programme effervescent and lively. My role is to give the mathematical perspective - but how this is done has changed over the three years of the programme. In 1986 the puzzlers solved the puzzle, with or without assistance, and then I was brought in to explain the mathematics that they had used or to provide another solution. This sometimes worked, but sometimes it was very evident that the puzzlers had not applied mathematics at all and moreover, having finished the puzzle, did not want to be told another 'more elegant' strategy.

Now my role is to give a hint part way through the action: to try to highlight a particular part of the puzzle which is significant and to bring in a new perspective which I think will help. I suppose I try to give the puzzlers and the audience a mathematical lens, through which they can look at the puzzle and then, with this new view, see ways to solution. We want to try to capture the real thrill of suddenly 'seeing it'. My hint has four aspects:

* What is difficult about the puzzle? What might be the problems for solution?
* What mathematics do I know that would seem relevant to the difficulty? What abstract thoughts or relationships or theories might it be possible to bring to bear?
* What do I know about these abstract thoughts? What deductive or formal reasoning can I use in this mathematical plane?
* How far does this help when applied back to the puzzle?

So to summarise, my role is to try to pin-point what the puzzlers might have found difficult, to try to introduce mathematical ideas, work with them a little in the discourse of mathematics and finally apply what I have done to the puzzle. In the 1989 series we have also introduced another new feature where sometimes, if the puzzlers have finished the puzzles successfully, I just comment on it very briefly and then offer an extension that people might like to think about at home.

3. A PUZZLE

To give a flavour of Fun and Games I will now describe an example of one of the puzzles. I will describe it in historical terms; that is in terms of the development of a puzzle from the initial idea to its eventual transmission. This puzzle first came to our attention during the 1987 series where it was found in several books and in various guises. The most lurid came from a fairly old source which went something like the following:

> Eighteen nuns live in the residential block of a certain nunnery consisting of eight cells which, together with the central courtyard, form a 3×3 square. The Mother Superior has decreed that there should be six along each side of the block, and she comes to check them out at irregular intervals both day and night.
>
> How do the 18 nuns fit into the cells so that there are 6 along each row?
>
> One night, after her evening inspection, six of the nuns slip out for reasons that need not concern us. How can the others so arrange themselves that the Mother Superior notices nothing amiss?
>
> Later, the six return, each bringing a companion. Is it possible for all 24 to so disport themselves as to obey the regulations?

We all liked the puzzle because it was challenging and demanded thought. It could also readily lend itself to activity with props which would both be helpful in the solution of the puzzle and therefore valid from a mathematical perspective but also would give the puzzlers something to do and the viewers something to watch. This sort of activity gives time and space for everybody to think about the puzzle.

However in 1987 the puzzle was rejected on the grounds that it was too complicated, in two ways. First, it was too complicated mathematically as there were too many distinct solutions. Second, it was too complicated visually as there were too many nuns. If we were to put eighteen objects on the screen the viewer would simply not be able to count them all. In 1989 we returned to the puzzle and tried to simplify it in both these two aspects. We felt that the puzzle could still be fun for smaller numbers. The number nine seemed to be the smallest number of objects for which it remains a puzzle. With only seven objects it becomes too easy.

Guards were chosen instead of nuns probably because they are less contentious and also the producers could imagine Rob having fun drilling the guards! Once guards had been decided, the producers then had to think about how big to make the props. In a book the puzzle is a diagram and the size does not matter. In Fun and Games the producers always favour big props that can be moved about. The first idea was to ask members of the audience to act as guards. This was rejected as asking people to arrange themselves would take too long and also would not help the puzzlers and the viewers to see any patterns. Apparently the producers seriously considered inviting in the Grenadiers along with a sergeant to drill them! This no doubt would have been great fun but would have made the puzzle rather opaque! One of the cardinal principles of Fun and Games is that the props should not get in the way of the puzzle - however tempting this might be! The puzzles are abstract and the props are concrete and they must not conflict.

We eventually came to the conclusion that the studio audience who are between five and ten metres from the action must be able to see what is going on and this would mean that the props should have dimensions of 20/30 cms. The guards could also then be handled but not hidden by the hands of the puzzlers. The guards were set in a wooden frame 3 x 3, as shown in the figure below (the cells were not labelled).

        A B C
        D (E) F
        G H I

At the beginning of the puzzle all the nine guards were in the space E and the puzzlers had to arrange them in the other eight slots so that there were *four* guards on each side of the block. On transmission Rob invited the puzzlers to have a go and after a couple of abortive attempts, I was brought in to give a hint.

Obviously there are a great many hints that could be given here and probably the one that most immediately comes to mind is to calculate the total number of guards that must be in the corners as they will be counted twice in the sum of all the rows and columns. In this case, since this sum is 16 and we have nine guards, we have to have seven guards distributed amongst the four corners. However, after much deliberation, taking into account the fact that the hint has to be precise, clearly given in a very short period of time and visual, I gave the following hint: if we have four guards in line A B C and four guards in line G H I, making eight guards, this means there is one guard in either cell D or cell F and none in the other since there are only nine guards altogether. A similar argument can be applied to vertical lines A D G and C F I. Both sets of lines were shown by gesture on the television - there were no labels. After the hint, the puzzlers solved the problem reasonably quickly.

When we watched the puzzle on the television, we were happy with the hint. With hindsight, the guardsmen should have perhaps been a little bit fatter or dressed in brightly coloured hats so that they would have shown up more clearly in the overhead camera. Also it might have been better to mark a red

Mathematics in Prime-Time Television 131

strip across cells A B C and G H I and a blue one down A D G and C F I so that in my hint I could have mentioned the four guards in the two red rows which would have been very clear and more visual.

As with a great many puzzles in Fun and Games, the guards puzzle can be a starting point for more mathematical investigations: for example, how many different solutions are there? What about if you change the number of guards? How do you work out a strategy for solution? How many solutions are there in each case? Is there a general solution to find all possible configurations for a given number of guards and to work out the number of distinct configurations

4. SOME RESPONSES TO THE PROGRAMME

In addition to good viewing figures, the response of the press to Fun and Games has been largely positive. In the Daily Mail for example, Friday 28 April 1989, Philip Purser said "Beneath its relentless high spirits Fun and Games (ITV) is intellectually unique among games shows and quizzes. It demands answers actually worked out by mental agility rather than by recall. In theory, a bright five year old and a law lord stand an equal chance of winning since acquired knowledge and wisdom play no part". This may not be true but it is interesting to read!

Some reporters expressed surprise at the success of Fun and Games, for example Northants Evening Telegraph under the heading "Calling all Clever Cloggs" wrote on 27 April 1989, "Television doctor and clever cloggs Celia Hoyles team up again tonight to present another programme packed with puzzles, conundrums, brain-teasers and games. Fun and Games is the show that aims to shake the cobwebs out of mathematics. But probably the biggest puzzle is how it has managed to capture the imagination of the public".

Obviously, although the majority of press reports have been complimentary, some have not. Most of the criticisms are interesting since they seem to stem from the standard stereotypical view of mathematics. For example, in the Sunday Express April 30 1989, it was suggested that I was "expected to recite twenty things you never knew about an isosceles triangle!" Some viewers also objected to the fact that at prime-time television you were expected to think a little!

This point of view was actually contradicted by the Yorkshire Post in a piece on Friday April 28 1989. Eric Roberts wrote that he felt, in the past, television had been "guilty of underestimating the concentration level of viewers". He noted that people were joining in the problem-solving in Fun and Games, both during the programme and at home. They loved the challenge of the puzzles. The same reviewer also wrote that there was a welcome absence in Fun and Games of "brass bands and the off-screen shouting which characterised and ruined so many other shows". What I think Mr Roberts is hinting at is that Fun and Games tries to tap and build upon the intuitive talents of the population, in, we hope, a way that is not patronising. The audience is not rehearsed and the puzzles are not rehearsed by the puzzlers. There are some gloriously funny moments as well as some gloriously stimulating mathematical insights.

5. SOME CONSTRAINTS ON PRESENTING MATHEMATICS ON POPULAR TELEVISION

So what are *the major constraints* of putting mathematics on prime-time television? The *first* that immediately comes to mind is the stringent time constraint. Each Fun and Games programme is twenty five minutes five seconds long and, on an average, we have five puzzles per show. So, roughly, each puzzle takes about four minutes. In that short space of time, the puzzle has to be stated clearly, the mathematical hint has to be given, and then hopefully the puzzle brought to a triumphant conclusion. That is just the mathematics - there are jokes and interactions between presenters and puzzlers which all bring the show alive. So there is enormous time pressure.

So why is all this pace necessary? From my perspective, I would prefer to spend longer on fewer puzzles but I am assured we would lose our audience. There is therefore a danger that Rob and I drive the puzzlers through to a successful solution rather than give them space to puzzle things out for themselves. We have to keep the puzzle moving - but not too fast! This is the tension we feel all the time on the studio floor.

We cannot expect viewers to grapple at length there and then with all the mathematical ideas they bump up against in the puzzles. All we can hope for is to leave the viewers with just a few ideas, or an intriguing idea that they can struggle with later at their leisure. In fact we do have a lot of telephone calls and letters from people who do just that - from mathematicians to members of the public who boast their previous loathing of mathematics. Some letters sadly do not refer to mathematics and just say, for example, that "they like my clothes" but the majority do give new solutions or extensions to our puzzles, often including ideas we have never thought of ourselves. One example of a puzzle that triggered quite a large correspondence was the Riffle-Shuffle Puzzle which was transmitted in 1988. Here we had a croupier who was doing 'perfect' riffle-shuffles; that is cutting the pack in half and then shuffling the cards so they went left, right, left, right, in perfect order. The puzzle was: how many ruffle-shuffles were needed to bring the cards back to the same order that they were at the beginning? People wrote computer programmes, people generalized from 52 cards to n cards, children undertook investigations in school and sent in their findings all around this puzzle!

The *second constraint* on the presentation of the mathematics in Fun and Games is that we have to make a very conscious effort not to appear to be teaching mathematics or to use any medium that calls up school mathematics. Sometimes it would seem to me to be obvious to have a diagram or, dare I say, a blackboard but we have to avoid this. We want people to break free of their past school experiences.

A *third constraint* is on the selection of puzzles. We try to use a variety of types of puzzle, some topological, some number manipulation, some more geometric, some games. All the puzzles though have to involve the puzzlers in doing something active. It is very bad television to see people just standing thinking. This obviously is a constraint because mathematics requires reflective thought. However, we do not expect people to become instant mathematicians! We just hope that we trigger an idea in their minds upon which they can reflect later.

The *final constraint* is that we do not use any formal mathematical language. This naturally restricts potential generality and power - but can release intuitions. I have found myself that applying formal symbolisation to the solution of a puzzle frequently is a barrier to adopting a more intuitive, commonsense approach.

6. CONCLUSION

In conclusion, I believe that Fun and Games has achieved at least some of its objectives. I believe that people are getting involved in mathematics although how far they are thinking mathematically, we will never know. Some people, though, do say they think differently about mathematics and mathematicians as a result of following the series. In a questionnaire survey after the 1989 series, respondents were asked, "if having seen Fun and Games they had changed their view of mathematics". Yorkshire Television were a little disappointed that only 20% said 'yes'. I was absolutely delighted. If we had made an impression on say 20% of nine million people we were not doing badly in popularising mathematics!

So perhaps it is now feasible to imagine a party when somebody says, in answer to the question "What do you do?", "I'm a mathematician" and instead of the usual "Oh, I could never do mathematics", or "I always hated mathematics", or deadly silence because nothing can be said, the answer comes up "Oh, how interesting, do you know this puzzle? 'If you've got a milk crate with 25 places in it, how many ways can you arrange 10 bottles so that there are never more than 2 bottles in any one line (horizontal, vertical or diagonal)?'" Wouldn't that be wonderful?

May I end by saying that mathematics is obviously much more than puzzles. Mathematics is a discipline with its own language and structures. Learning mathematics is also much more than watching Fun and Games or even doing the puzzles. Learning mathematics requires hard study, considerable thought and persistence. But if we want to popularise mathematics and open up the mathematical culture to a wider range of participants, first of all surely we must open people's eyes to what mathematics could be all about?

REFERENCES

[1] Dallas D. (1988/89) Television by Numbers: Does Science Programming have to be Popular? Airwaves, Winter, p.16.

[2] Fennema E (ed). (1985) Explaining Sex-Related Differences in Mathematics: Theoretical Models, Educational Studies in Mathematics, 14, 3, p.303-321.

[3] Hoyles C. (1985) Culture and Computers in the Mathematics Classroom, Bedford Way Paper, Institute of Education, University of London.

Cultural Alienation and Mathematics

GORDON KNIGHT

Massey University, Palmerston North, New Zealand

1. INTRODUCTION

The aim of the popularisation of mathematics is to influence the perception which people have of the subject. Since this perception differs in different sections of the community, it is firstly important to identify a target audience and then to seek to understand the nature and origins of the views which they have of mathematics. Blanket attempts at popularisation based on the perceptions which mathematicians have of their subject are unlikely to succeed.

In this paper, some factors associated with the popularisation of mathematics among the Maori people, the original inhabitants of New Zealand, are considered. It is likely that similar, but not necessarily the same, factors will pertain to other ethnic minorities with no strong formal mathematical tradition.

It will be argued that the Maori people have been culturally alienated from mathematics and that attempts to overcome this must go beyond the superficial introduction of elements of Maori culture into a traditional presentation of mathematics. Initiatives, by the Maori themselves, firmly based in their own culture have much more potential.

2. BACKGROUND

The Maori have been in New Zealand for about 1000 years. It is agreed that they derive from those Polynesians who first settled in East Polynesia. About 200 years ago the first European contacts were made, initially through explorers, scalers, whalers and missionaries, and then through systematic settlement from Britain from about 1840. Before this European contact the Maori had a stable and coherent culture and lifestyle.

This Maori culture had no written language and, as with many, if not all, oral cultures, the ethnomathematics of the culture did not have very much in common with Western formal mathematical tradition. With the arrival of the traders, however, the Maori soon recognised the need for the mathematics of the introduced culture and proved to be able pupils. One of the early missionaries (Duncan, 1853) comments on their aptitude:

> "many of them thoroughly understood the simple rules of arithmetic and could calculate readily".

This ability was not universally welcomed by the traders, some of whom suggested that the missionaries should restrict their teaching to religious matters. Their motives are clear, it was becoming more difficult to cheat the Maori. This is perhaps a simple illustration firstly of the empowering nature of mathematical knowledge and secondly of an attempt by people of one culture to deny that power to people of another.

It might be argued that where the intentional attempts of the traders to deny mathematical knowledge to the Maori failed, the domination of the Maori culture by the introduced culture has succeeded, largely unintentionally.

The problem is most evident in New Zealand schools. The country prides itself on its egalitarian society and on its equality of educational opportunity. There are certainly no legal barriers to Maori access to education. However, the education system has not been successful in catering for the needs of the Maori population. Evidence of this is provided by the national School Certificate Examination available to students at about 15 years of age. This examination plays a critical role in New Zealand education. It opens, or closes the way to opportunities in higher education and, particularly in mathematics, has a very significant role in giving or denying access to employment. New Zealand Department of Education statistics show that a non-Maori student entering secondary school is of the order of 3.5 times more likely than a Maori student to leave school with an acceptable grade in School Certificate mathematics.

It would be a mistake to attribute all of this difference directly to ethnicity, there are other factors involved such as socio-economic status. However, Garden (1984) found in a study of ethnic factors in the New Zealand data from the IEA Second Study of Mathematics that there was a significant difference in performance which he could only attribute to ethnicity. The social and political implications of this situation are obvious and are evident in the situation of other ethnic minorities around the world.

3. CULTURAL ALIENATION AND MATHEMATICS

The results of the failure of the New Zealand education system to cater successfully for Maori students and to capitalise on the Maori aptitude for mathematics which the missionaries found are deep seated and no easy remedy is likely to be found. An important first step, however, is to acknowledge this mismanagement and to recognise that one of its major effects has been that Maori students feel culturally alienated from mathematics.

Mathematics is, and for the last 150 years has been, taught almost entirely by non-Maori teachers, in English, using syllabuses and textbooks reflecting the dominant non-Maori culture, in institutions whose structures and values take no account of the structures and values of Maori society. It is little wonder that generations of Maori students have come to regard mathematics, along with other school subjects such as science (Stead, 1984) as 'pakeha' or non-Maori subjects.

This cultural perception then interacts with any cognitive difficulties in a distressing way not dissimilar to the interaction of cognitive and affective factors associated with mathematical anxiety. When Maori students are not succeeding, whether or not this is due to cultural factors, their alienated view of mathematics is reinforced. Why should they expect, or even want, to succeed in this 'pakeha' subject. In attribution theory terms we have a clear case of 'learned helplessness'.

The major problem of the popularisation of mathematics with the Maori is one of overcoming this cultural alienation.

4. A MAORI PERSPECTIVE ON MATHEMATICS

For an educator with a western cultural background, the obvious way to tackle this situation is to introduce elements of Maori culture into the presentation of mathematics. Such an approach has the label 'taha Maori' (Maori perspective) in New Zealand. It has included trying to eliminate cultural bias in textbooks by including Maori people in illustrations, using Maori names for people in word problems, and even numbering the pages with Maori numbers. This has many parallels with attempts to eliminate gender bias in mathematics teaching. Opportunities have also been taken to introduce Maori examples in the content. Polynesian navigation methods may be discussed and elements of Maori art are introduced into geometry lessons.

Such initiatives are, of course, well intentioned and similar approaches have been tried in other parts of the world. Unfortunately, in New Zealand at least, they do not seem to work. In fact they seem to benefit non-Maori students more than Maori and almost certainly fail to really address the problem of cultural alienation. For non-Maori students the approach gives them another perspective on their already secure view of mathematics and may serve to promote cultural awareness. Many Maori, however, reject the initiatives as 'tokenism' and some regard the approach as positively dangerous since it salves the conscience of the pakeha without confronting the real issues.

In a study of the attitudes of Maori students to science, Stead (1984) argues that the rejection of 'pakeha knowledge' by the Maori is a reaction to the rejection by the dominant culture of New Zealand of the knowledge and values considered important by the Maori. This view is supported by Cummins (1986) who stresses the importance of status and power relationships in any account of minority group performance in education.

If this is the case, in order to popularise mathematics with the Maori it must be presented, not as 'pakeha knowledge' to which elements of Maori culture have been attached, but as 'Maori knowledge'. Ideally, this might involve the same techniques which gave the subject its current cultural image. Maori teachers teaching mathematics in the Maori language, using syllabuses and resources which reflect Maori culture in institutions whose structures and values are those of Maori society. Although far from this ideal, current initiatives by the Maori people themselves are very much along these lines and have considerable potential for success.

5. MAORI STATUS AND PRESTIGE IN RELATION TO MATHEMATICS

The fundamental principle of these Maori initiatives is that instead of starting with mathematics and introducing a Maori perspective, once must begin with Maori culture and introduce a mathematical perspective. In this way, the status, power and prestige, the 'mana', is given to Maori rather than to mathematics. There is no other way in which mathematics can be accepted as 'Maori knowledge' and the cultural alienation be overcome.

This focus on Maori culture rather than mathematics as the origin of any popularisation enterprise applies at least as much to the context and process of presenting mathematics as it does to the mathematical content.

Cultural Alienation and Mathematics

At the very heart of Maori culture is the Maori language.

> 'Ko te reo te mauri o te mana Maori'
> (The language is the life principle of Maori power and prestige)

For many years after the arrival of the European, the use of the Maori language was very strongly discouraged. It was assumed that it was in the best interests of the Maori to forsake their old ways and to learn the ways and values of the colonists. Many children were punished for speaking the Maori language at school. As a result of these assimilation policies the language was in serious danger of being lost altogether. In recent times, however, there has been a renaissance of cultural awareness by the Maori centering on a revitalisation of the language. Increasingly Maori is being taught in schools and being spoken on radio, television and even in the courts.

One of the most important features of this renaissance has been the establishment, by the Maori people themselves, of a pre-school total immersion language programme called Te Kohanga Reo (the language nest). The success of this enterprise has led to an increasing number of bilingual programmes in primary and secondary schools. It is within these programmes that the potential for dealing with the cultural alienation which Maori students experience in relation to mathematics exists. At secondary school, bilingual programmes have been operating only for a very short time and the mathematics teaching within them is hampered by a lack of Maori speaking teachers. There are also difficulties associated with the constraints of working within an examination dominated system when the examination reflects the knowledge and structures of the dominant society.

In spite of these difficulties, a visit to any one of these bilingual units will indicate just how powerful a change of context of this kind can be in influencing students' attitudes to school in general and to mathematics in particular. Many of the students find for the first time that their Maoriness can be an advantage, not a disadvantage, in learning mathematics and that mathematics can be enjoyable. This is surely an indication of the reversal of cultural alienation. Research by Wagemaker (1988) also indicates that the bilingual approach is increasing retention rates at school for Maori students.

The constraints of examination prescriptions make changes in content more difficult to achieve and there is still a good deal of work to be done here. A proposed relaxation of these constraints should help considerably. The most promising approach to content is described in Begg (1988) and involves a theme based approach in which topics are taken from the Maori language syllabus and the mathematics associated with these topics is studied alongside the language study.

For example, one of the language topics is 'Kai' (food). In the language class the students discuss favourite Maori foods, looking for and growing foods, preparing and cooking food, and the Maori custom of feasting. In the mathematics class they might discuss budgeting (arithmetic), volume, weight, cooking temperature etc (measurement), growth curves for plants and animals, cooking times etc (algebra), analysis of foods, sampling (statistics) and shapes of food and packaging (geometry).

This approach to learning which blurs the boundaries between subject areas is very much in line with the traditional Maori ways of learning.

6. CONCLUSIONS

There are a number of general lessons for the popularisation of mathematics among culturally alienated people which can be drawn from the New Zealand experience which has been described.

Firstly, the initiative for the changes came from the Maori people themselves. Sympathetic non-Maori mathematicians have been able to contribute, but it is vital that they stay in a supporting role. Any bid by the dominant culture to take over and control the enterprise would change the critical status relationship of the enterprise and render it ineffectual.

Secondly, not only are status relationships between people important but also those between values and different kinds of knowledge. Mathematics must take the subordinate role and fit into Maori culture, not the other way round. This has implications for both the content and the means of transmission. The content must come from the culture and be transmitted through the culture if the target audience are to identify with it as 'their' mathematics.

Thirdly, the New Zealand experience provides support for the view expressed by Bishop (1988) that mathematicians confuse the 'universality of truth' of mathematical ideas with the cultural basis of mathematical knowledge and expression. The solution to the problem of cultural alienation from mathematics is totally dependent on the acceptance of this view.

In response to their desires to once again be involved and competent in mathematics, the Maori people are pioneering a new relationship between mathematics and culture which has implications for other subject areas and other cultures.

REFERENCES

[1] Begg A J C. (1988) Mathematics, Maori Language and Culture. Paper presented to ICME 6, Department of Education, Wellington.

[2] Bishop A J. (1988) Mathematics Education in its Cultural Context. Educational Studies in Mathematics, 19:2, 179-191.

[3] Cummins J. (1986) Empowering Minority Students: A Framework for Intervention. Harvard Educational Review, 56:1, 18-36.

[4] Duncan J. (1953) Progress of the New Zealanders in Civilisation and Religion No IV. The Scottish Presbyterian, Nov 1853, 331-338.

[5] Stead K E. (1984) An Exploration of Different Outlooks on Science: Towards an Understanding of the Under-representation of Girls, and of Maori and Pacific Island Students in Science. PhD Thesis, Waikato University, Hamilton.

[6] Wagemaker H. (1099) Maori-English Bilingual Education: Tauranga Boys' College. Research and Statistics Division Research Report Series No 51, Department of Education, Wellington.

Solving the Problem of Popularizing Mathematics Through Problems

MOGENS ESROM LARSEN

Københavns Universitets Matematiske Institut, Universitetsparken 5, DK-2100 København Ø, Denmark.

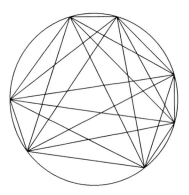

fig. 1

During the last three years I have been constructing puzzles for the popular science magazine *Illustreret Videnskab*, published monthly in Denmark, Finland, France, Norway and Sweden. I believe that for many people it is pure fun to solve mathematical problems, if the task is voluntary, and the problems challenging. Hence supplying a popular magazine with entertaining problems is a great chance to lure mathematical methods of thinking into the minds of the readers, even if they never appreciate this aspect of their behaviour.

I think from my own experience as problem solver and poser, that it is wiser not to over-estimate the reader's knowledge, and not to under-estimate their

Popularizing Through Problems

intelligence. So I hesitate to ask too stupid questions, but not to ask difficult questions which are easily understood.

The puzzles are mainly classical ones from H. E. Dudeney, Sam Loyd etc., but I try to sneak in a little mathematics here and there. E.g. in the problem of the jeep crossing the desert, I added the question, "how big a desert can we cross?" I hoped that some readers would prove the divergence of the harmonic series, and according to letters some did.

To tease computer-freaks, I like to ask questions with very large solutions. A source to such problems is Pell's equation. Some of these problems I have discussed in [2], but another is the following:

"The pride of the republic of Inner Urdistan was the army. Each year every one of the 60 regiments sent 16 soldiers to the parade. They marched in 60 squares. Then the general, M. Urder, joined the forces and all of them formed one big square together.

"After the revolution, the leaders founded a new regiment making a total of 61. But the general wanted to form the parade all the same. So he asked the 61 regiments each to send a square number of soldiers such that they could all together including himself form a new big square.

"How many soldiers did each regiment send?"

In this case we must solve the equation

$$x^2 - 61 \cdot y^2 = 1$$

The smallest solution is

$$51145622669840400.$$

Among the numbers up to 100, the biggest solution is required by 61.

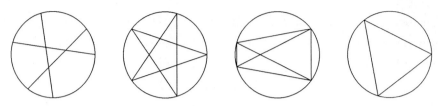

fig. 2

Counting is good, but formulas are better. One example is to draw a heptagon with all diagonals and count its 287 triangles. (See figure 1.) Another is to find the number of triangles in an n-gon in general position with vertices on a circle, see [3], the solution is

$$\binom{n}{6} + 5\binom{n}{5} + 4\binom{n}{4} + \binom{n}{3}$$

to be found in [1].

As stressed by Richard K. Guy in his answer to [1], this formula becomes evident if we look at figure 2 carefully.

It is even more fun to count the number of triangles in a regular triangular lattice, see figure 3; an old reference is [5], but see [4] on the general formula:

$$\left[\frac{n(n+2)(2n+1)}{8}\right].$$

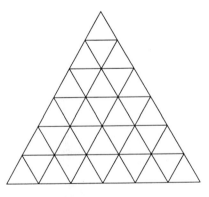

fig. 3

Theorems are still better than formulas. Take a paper with the ordinary square lattice. Take any 5 lattice points and prove that there must be two of those 5 points, such that the line segment from the one to the other goes through a lattice point. (The pigeon-hole principle)

As computation can give the security of a result, and this is second to the understanding obtained from the logical deduction of a theorem, the latter is second to the *insight*, the experience of seeing right through the problem as can happen in geometry. Draw two intersecting circles and ask for that line through one of their points of intersection which is longest: in figure 4 we ask for the choice of line through A, such that the segment BC is longest.

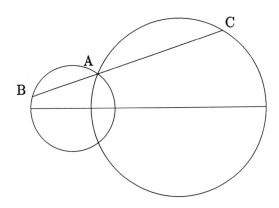

fig. 4

The solution is surprisingly simple. Draw the triangle $\triangle BCD$, where D is the other cutting–point of the circles. See figure 5.

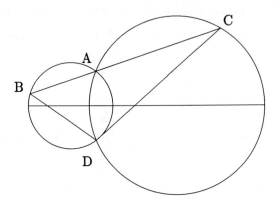

fig. 5

Then it is obvious, that the angles at *B* and *C* are independent of the particular choice of line through *A*. All the different triangles are similar. This means that we obtain a maximal distance *BC* if we choose a maximal distance *DC* (or *BD*). And this is easy, we have to choose one (and then both) as the diameter from *D*. See figure 6.

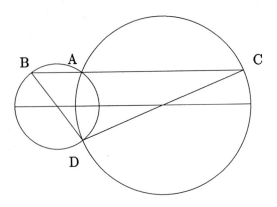

fig. 6

Then *BC* becomes parallel to the centre-line.

Popularizing Through Problems

Even more fun than geometry will be topology. Everybody knows the impossibility of joining three utilities with three houses in the plane without crossings. That is to say, the complete graph between two sets of three elements is not a planar one. But if we change the conditions to a non-planar surface, e.g. a torus, a Möbius band, or the projective plane, then it might be solvable. And this is seen in figure 7 below.

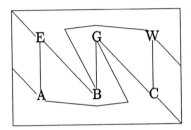

fig. 7

This solution works equally well whichever one of the interpretations of therectangle you prefer.

These are examples of what I like to do. Of course, it is a problem to continue the supply. I have published about 500 problems over 3 years; how can I do that? The easy answer is that I feel free to steal problems everywhere. The history of the problems shows that stealing them has been the custom all the time. So, I have a couple of problem collections and a couple of friends, with whom I exchange problems.

Whether my efforts are worth-while or not is hard to tell. I know from myself how much I enjoyed doing such problems as a youngster, so I naïvely assume that others will enjoy the same. But I know from teaching friends that a problem I have posed is immediately challenging, so that all the pupils want to see it solved. In this way my problems have made it easier to get a class started on integration or whatever they must. And if I have contributed to making the teaching of mathematics just a little more entertaining, I find it worthwhile doing so.

References

1. Cormier R J and Eggleton R B. (1976) Counting by correspondence, *Math. Mag.* 49, pp.181-186.

2. Larsen M E. (1987) Pell's equation: a tool for the puzzle-smith, *Math. Gazette* 71, pp.261-265.

3. Larsen M E. (1988) Problem 72A, *Math. Gazette* 72, p53.

4. Larsen M E. (1989) The eternal triangle, *College Math. J.* November.

5. Loyd S. (1914) *Sam Loyd's Cyclopedia of 5,000 Puzzles, Tricks and Conundrums with Answers*, Corwin Books, New York pp.284,378.

Popularizing Mathematics at the Undergraduate Level

BRIAN MORTIMER AND JOHN POLAND

Department of Mathematics and Statistics, Carleton University,
Ottawa, Canada

This is a personal account of an attempt, and failure, to popularize mathematics in the undergraduate community of a university. We hope to convince you that this was an exciting adventure, with much scope and promise, and to warn you of the political problems that arose and finally aborted the project. The idea was to offer a mathematics enrichment course for students who had completed freshman mathematics but were pursuing careers in other disciplines. Two particular target groups were in mind: the first consisted of students and alumni we meet occasionally who are still enthusiastic about mathematics. Although they never went beyond freshman mathematics they can still wax nostalgic about their former romance with mathematics. They want to know more about mathematics. The second was a more nebulous group, students who might be destined for positions of influence for funding mathematics at the national level, and therefore presumably would obtain university degrees and see something of mathematics along the way.

To the first group we hoped to offer a course that was exciting, satisfying and gave a broad perspective on the role of mathematics. Mathematical in nature, but naturally not a standard, technical mathematics course. To the second group, we hoped to show the centrality of mathematics, both historically and presently, and to convey a picture of a living, evolving subject with many questions and issues yet unresolved.

The catalyst for the course came three years ago, when it was mathematics' turn to choose a speaker for the annual science lecture at our university.

Sir Michael Atiyah accepted our invitation, and proposed to speak on "Geometry and Physics: Euclid, Einstein and Elementary Particles"[1]. The auditorium was unexpectedly packed, with people sitting in the aisles (my wife arrived at the last moment and was turned away for lack of space). In previous years, talks in other areas of science drew slightly better than half this size. A mathematics talk outdrawing talks in computing, bio-engineering, or nuclear physics? How unexpected; how delightful! Flushed with such success, we sat down to draw up a course outline based on the topic of geometry and physics. It would trace the evolution of mankind's ideas on the geometry of the universe.

Here is a description of the course and an outline of the thirteen weeks it would be offered (three hours per week).

Week 1: Euclidean Geometry: Pythagoras; Unique parallels; sum of angles in a triangle; locus equidistant to a line; Greek outlook on the cosmos, history and attitudes of Greek geometry.

Week 2: Spherical Geometry: bounded lengths, bounded areas, angles and lengths equivalent; no parallels, no similarity; spherical trigonometry and navigation; history of spherical geometry, non-acceptance as a Non-Euclidean geometry.

Week 3&4: Analytic Geometry: Descartes and higher dimensions, concept of distances; cosmology of Kepler and Newton, elliptic orbits; "The Elements" and the "Principia".

Week 5&6: Hyperbolic Geometry The Poincaré model; many parallels, no similarity; angles and length unique natural length element. History of the parallel postulate: development of facts and attitudes. Three dimensional hyperbolic space.

---

[1] a copy of the poster (designed by mathematicians) advertising this talk appears in the appendix

Week 7&8: Minkowskian Geometry: New metrics on R4; constant speed of light and other background to special relativity and space-time; the paradoxes. Relation of Minkowskian Geometry to Hyperbolic Geometry.

Week 9,10,11: Manifolds: Definition; local vs. global properties; scalar, vector and tensor fields; the metric tensor, relation to classical geometries. Curvature: extrinsic and intrinsic; the curvature tensor.

Weeks 12: Geometry of the Universe: Interpretation of a force as curving space-time; energy-momentum tensor; Einstein equations. Higher dimensional manifold to accommodate all forces of nature; singularities in space-time; black holes.

References

Friedricks K O. (1965) *From Pythagoras to Einstein*, New Math. Lib. Singer.

Gray J. (1979) *Ideas of Space: Euclidean, Non-Euclidean and Relativistic* Clarendon, Oxford.

Greenby M J. (1980) *Euclidean and Non-Euclidean Geometries; Development and History* W H. Freeman, 2nd Ed.

Kline M. (1953) *Mathematics in Western Culture* Oxford, U.P.

Lanczos C. (1970) *Space Through the Ages: The Evolution of Geometrical Ideas from Pythagoras to Hilbert and Einstein* Academic Press.

Penrose R. (1980) "The Geometry of the Universe", in *Mathematics today*, ed. L.A. Steen, Vintage.

Schutz B F. (1980) *Geometrical Methods of Mathematical Physics* Cambridge U.P

Yaglom I. (1979) *A simple Non-Euclidean Geometry and its Physical Basis* Springer.

Texts

The book by Lanczos covers all the ideas of the course. It may be heavy going for a student with the bare prerequisites, but would be suitable for the instructor and some students. The book by Gray covers essentially the same material and would be useful to all students but would not have a high enough mathematics content. We would like a book halfway between. This could be achieved by taking three of the books in the reference list (Greenberg, Penrose and Fredericks) but not in any one yet found.

As our enthusiasm grew for the subject matter of this course, we came to see how ideal it was as a vehicle for popularizing mathematics. Geometry of the universe begins with the most approachable of topics - geometry - and links it with a higher-order desire, our need to understand and have a general picture of the universe in which we live. Surely, this is the foremost task of a popularization course: to be able to show the importance and relevance of some area of mathematics in the science and culture of humanity, and yet at the same time maintain some easy familiarity and accessibility for the student. For example, it seems unwise to rely heavily upon algebra, a less friendly topic for most, nor should it attempt to aim low, at simply explaining how some technological advance depends upon mathematics. In these aspects, the geometry of the universe is more apt then, say, number theory and cryptography, just to take one obvious alternative. Students at university, despite dry academic courses, really are looking to make their way in a realm of deep questions, questions of identity, relationships, career, and their place in the universe. They recognize the uncovering of the geometry of the universe as a serious, deep, and legitimate problem. Many of our colleagues, upon first hearing of our proposal agreed immediately that this was an important subject matter that we ought to be offering.

The evolution of geometry historically has had many significant interactions with philosophy, culture and how we view the world (see for example

J. Grabiner [1] or M. Kline [2]). A major side-effect of this course would be to expose the student to this rich context in which the geometry developed, adding enormously to the meaning derived from this course. Tracing this evolution will also make it clear that our present-day models are likely not the final ones, and that our quest continues. The audience must be left with the impression that mathematics research is a lively and vital area of human activity. Notice that the major thrust of this course is not the rationalization of the importance of the milestones of mathematical development, such as projective geometry or the calculus. Certainly these would be major biproducts, but the students' eyes would be looking higher, at the quest to explain our universe. And that makes a perfect vehicle for these other objectives, of showing the centrality, relevance and vibrancy of mathematics.

There is a very delicate problem here of how this course should be taught. The purpose is to expose students in some forty hours of lectures to a series of increasingly sophisticated mathematical ideas (such as hyperbolic and Minkowskian geometry, curvature, manifolds, and present-day models), with an emphasis upon leading them up to an overview of the current issues. As background preparation, we assume only freshman calculus and linear algebra: a tenuous familiarity with differentiation, integration, Euclidean space and complex numbers. (As an aside, we mention that this is required in all programs in science, engineering, and commerce so this is a reasonable requirement.) How can this gap be bridged? It is necessary to find keys that will bring advanced topics to such students and avoiding paths that would get ensnared in the material from intermediate-level mathematics course. Attempting to work with the standard logical development from carefully worded and irrefutably accurate definitions via nonexplanatory but slick proofs would spell disaster. The teacher for this course must see, confront and overcome the culturally-defined mystique about how university mathematics courses ought to be taught that prevails so widely today. Otherwise the popularization of mathematics at this level is doomed.

This would not be an easy course to teach. No textbook exists to our knowledge and even the reference books are not organized as textbooks, with carefully chosen exercises. In many ways, this is closer to a course in the Arts than a technology course. It deals with a central intellectual question, with a huge and relevant literature. The teacher must bring to bear almost all he/she knows in this vast area to be effective in teaching. Most challenging is this need to be prepared to work with a wide variety of deep relevant student questions that should arise while teaching each topic. And the students must be seen, and their immediate reactions, to the teaching used as a tool for developing the right keys to open advanced topics to them.

Is it possible? Can the gap be bridged between advanced topics and unsophisticated students? Hopefully, this is a major facet of what popularization of mathematics is all about. We have experience over ten years of a weekly seminar with high school students, and another with undergraduates, presenting them with exciting, sometimes unorthodox, but invariably advanced material from automata theory, number theory, set theory, geometries of many types, and dozens of other areas. Also, working on Expository Mathematics: an annotated bibliography (by John Poland, M.A.A., 1989, to appear) led to the discovery of expository articles at all levels of difficulty and varying from a few pages to book length: surely such variety indicates that ways exist for this course to be taught successfully. There is a skill, and it can be honed, to presenting this material legitimately to "underprepared" students.

So, you are convinced that this course looks excellent. Why then did it never get approved by our department?? For one, we must remember that the age profile of most mathematics departments, including ours, reveals a primarily conservative group, trained in classical mathematics departments, and who never have seen such a course before. We tried to be quite explicit, even in the description intended for the university course calendar, that the material was to be primarily descriptive in nature and aimed at non-mathematics majors. But we deliberately did not want to see this course mired down in the safe but less vital discussion of just pre-twentieth century geometry. The reaction of many of our colleagues was that the advanced material simply could not be treated adequately without much more preparation or more time spent in detail in the course. In reply to our reminder that this was intended to be descriptive, colleagues argued that it followed then that this course was not primarily mathematical in nature, but rather closer to the history of science. We replied (both to our mathematics colleagues and colleagues in history and philosophy) that the design of this course was to expose the student to the intellectual heritage that mathematics represents. It had as its prerequisites freshman calculus and linear algebra in order that the students could come to grips with some of the mathematics involved, but a delicate balance between mathematics and its wider effects would have to be struck throughout the course. In the end many of our colleagues in mathematics became convinced both that some of the topics were too advanced and that the course was not mathematical enough in nature. It would have to treat these topics superficially and that just was not how mathematics courses should be taught. Everyone realized that the course was dealing with central issues in civilization that ought to be taught, but at the same time it was important to many of our colleagues that this subject be dealt with properly and not just superficially. We had hit a central nerve, it seems!

Perhaps it is worth noting that more recently, in discussing possible changes to freshman calculus in our department, many colleagues argued that, acquiring technical facility is the keystone to successfully doing mathematics. Alternate approaches such as more conceptual questions, should only be addressed once this technical mastery has been demonstrated - otherwise the students in mathematics courses would be short changed.

Mathematics has ideas and techniques, and some mathematical ideas are embedded in technicalities while others stand alone. This is true in any discipline. It is possible to have a mature sophisticated understanding of some ideas without being in a position to use it creatively (Natural Selection in biology might be a familiar example). Our debate in the department was to some extent about whether a mathematics course can legitimately teach ideas without techniques. Do we have a tendency to not allow our students to listen to Beethoven quartets in class until they have demonstrated that they know how to compose; they cannot really appreciate the quartets or even understand them otherwise? What then is music appreciation?

The need to popularize mathematics to a mathematically-literate audience of non-mathematicians seems vitally important. A university course seems the natural vehicle (our university even offered the possibility of a televised lecture series locally). And what could have as compelling relevance, importance and abiding historical and cultural interest as the geometry of the universe?

Bibliography

[1] Grainier J V. (Oct. 1988) "The centrality of mathematics in the history of western thought", *M. Mag.* 61:4, p.220-230 (also in *Proc. Internat. Congress M.*, Andrew M Gleason, ed., A.M.S., 1987, p. 1668- 1681.

[2] Kline M. (1964) Mathematics in Western Culture, Oxford U. Press.

The Undergraduate Level

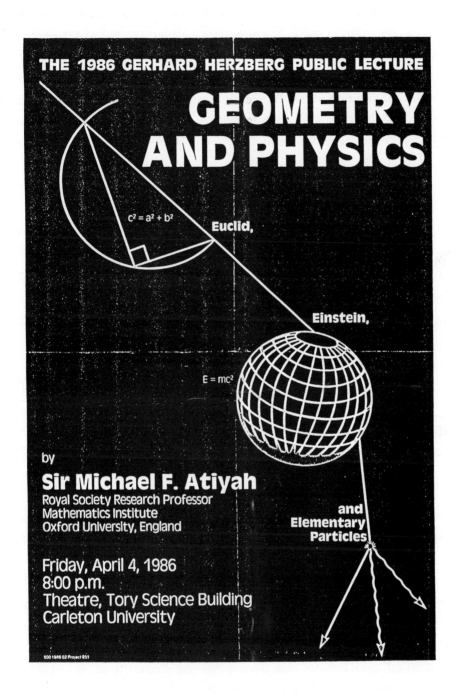

The Popularization of Mathematics in Hungary

TIBOR NEMETZ

Mathematical Institute of the Hungarian Academy of Science, P.O. Box 126, H-1364 Budapest, HUNGARY

1. INTRODUCTION

Let us start with a quotation from J. I. Ignatyev:

"Nowadays there is hardly anybody who would deny the necessity of the most wide-ranging popularization and diffusion of mathematical knowledge. Basic mathematics should be provided already at a very early age during the process of education and instruction. .... We may expect success only if we use every-day situations, familiar topics as keys to enter the territory of mathematics by the help of problems that are, at the same time, stimulating and full of wit." This quotation dates back to 1908 but it is in perfect harmony with the discussion document [2]. If it was necessary to popularize mathematics 80 years ago, it is more urgent now, since "there is now an increasing divergence between the advancement of science and the general understanding of the vast majority of human beings". There are many efforts all over the world, and it is high time to collect and exchange international experiences. In this communication we present Hungarian examples and offer a few comments.

The paper is organized as a loose collection of items. These items reflect the author's experiences during 25 years of efforts for the popularization of mathematics, including the last 8 years when he was a member of the mathematical presidium of the Society for the Diffusion of Sciences, with Hungarian abbreviation **TIT**. Experiences gathered as a member of the community of mathematics educators are also incorporated.

The discussion document [2] distinguishes three factors of the process of the popularization

- Whom: the intended population
- How: the media to be used
- What: the topics to be considered

We would like to add two more dimensions:

- Why: why this topics to this section of the public in this way
- Timing: match specific topics to the given time

We will discuss the following examples according to these five dimensions, although not necessarily in this order.

2. ITEM 1: CLUBS OF YOUNG MATHEMATICIANS

Whom: Clear-headed pupils of the upper primary grades, age 11 - 14.

Why: The 1948 Educational Law has institutionalized primary schools with 8 grades of obligatory education for Hungarian children. The unified, obligatory syllabuses are aimed at the masses. The subject matter in mathematics was unsuited to attract the attention of more talented children. Since it was hopeless to change the official school system, the problem had to be solved outside of it.

Timing: Political changes permitted such solutions by 1960. The work could be officially organized under the auspices of the Society for the Diffusion of Science, TIT.

What: anything which is suitable to raise interest at the given age.

How: Volunteers compiled work-sheets as frames for clubs called Clubs of Young Mathematicians. The clubs meet every second week during the school year and work under the direction of enthusiastic teachers. The clubs are nested all over the country with an average of 17 members and with a total of around 14.000 students yearly. The operating of the clubs is served by a nationwide competition. These competitions are described in [4].

3. ITEM 2: CONTEST BY CORRESPONDENCE

Why: There are open-minded children also in small villages and, in general, in rural communities. Their small number, and the lack of enthusiastic teachers may prevent the establishment of the clubs described in Item 1. Therefore the Society TIT has looked for other means for serving them.

The Whom and By What Topics are the same as before.

How: Several county organizations of the TIT have announced grade-dependent, all-year-round competitions by correspondence. They prepare 7 - 9 sets of problems ranging from easy to more difficult ones, and mail them to the applicants. All problems are assigned some points of difficulty and these points appear on the problem sheets. The children are asked to submit their solutions with reasoning. Every grade is taken care of by one or two teachers who check the solutions, collect and record the points of everybody and return the solutions with personal comments. The popularity of this contest lies here. It is rather unusual that a village child gets a letter, it is even more unusual if it comes from SOMEBODY, and includes such comments as THIS TIME YOU HAVE REALLY DONE A GOOD JOB. Of course, the participating teachers must be very enthusiastic to perform this manual work almost free of charge. The present hostile financial regulations may put an end to this movement which became popular not only in rural environments but also in the capital.

The contest is usually concentrated in the period between October and April with more or less equal time among the distributions.

4. ITEM 3: TV COMPETITIONS

Sticking to the same age group we report briefly on a TV competition which was launched during the late seventies. This is part of a programme of the Hungarian school-television. The programme was initiated to help in overcoming difficulties raised by the introduction of the "new maths" in 1974. Problems on the screen reflect the spirit of the new maths in their variety. Applicants are filtered before letting them to the screen. Not only the parents but also the schools are very proud when their children qualify for the competition. As a side result, these contests have proved to be helpful in changing typical dislike toward mathematics in the adult population. Sitting in front of the TV they have as much time to break a problem as the children. And the nature of the problems guarantees that they can arrive at the correct solution.

More details on this "Fortune favours the brave" competition can be found in [4].

5. ITEM 4: RADIO PROGRAMMES

- The nationwide introduction of the "New Maths" created a demand for information on behalf of parents who were not familiar with the subjects their children were learning. The teachers themselves had every-day-difficulties due to lack of experience. In order to help them to cope with these difficulties, the Hungarian Radio launched a programme under the title "The children learn New Maths". The programme soon became popular and is still running.

- In the view of the author, two other programmes of the Schoolradio were not so successful. The first was edited for school children under the title "let us solve it", the second aimed at secondary schools (age 15 - 18) discussing topics neglected by the official curriculum at a popular level. There are many possible reasons for the failure. Among them a common habit of listening to the radio just for "background music" prevents students from listening to something which needs attention.

6. ITEM 5: PERIODICALS

■ The role of popular papers has been observed as early as 1894, when the worlds oldest mathematical monthly for school children (KöMal = Közéepiskolai Matematikai Lapok, Secondary School Mathematical Journal) started its career. Its popularity is shown by the average number of 3 - 4 subscribers per class. Subscription for a year costs 85 Forints. For more information the reader is referred to [7].

■ In 1986 a new periodical was established for school children under the title "Hátter" (Background). Following Japanese experiences, separate issues have been prepared for the age - groups 13, 14, 15, 16, 17, and 18, with two numbers per school year. It aims at general popularization with columns in language, literature, history, geography, mathematics, physics, chemistry, biology, technology and fine arts. We try to characterize its content by listing the mathematical papers in the first number for the age group 13.

"The father of the computers - Neumann János - John von Neumann" written by Dénes Gábor with quotations and photos from G. Dantzig and G. Pólya.

"Let us play with numbers" by the present author written on the mathematics involved in the Bar - Kochba game (known also as the 21 questions).

"The first revolution in computing" by Márton Sain, author of a "best-seller" on the history of mathematics.

"In heads? Or by Computers?" by Tamás Varga.

The numbers are sold at a price of 75 Forints, slightly less than KöMaL for one year. This might be responsible for a relatively poor circulation.

The periodical "Teaching Mathematics" deals mainly with methodological questions for teaching age groups 11 - 18. More or less regularly it also publishes popular articles which are well received by the community of maths teachers. The choice of subjects, however, does not show systematic planning. It seems to be governed by a routine of whom could the editors convince to write an interesting paper at no fee. Production costs must be kept to a minimum in order to keep the price of 7 forints per number. Its circulation indicates that it is read by practically all secondary maths teachers and many upper primary ones.

The public in its widest form is targeted by the weekly periodical of the TIT under the title "Life and Science" (Élet es Tudomány). Thousands were mobilized by a mathematical puzzle column in it. Its popular articles are well received by both young and old. Although its price of 14.50 forints is a relatively cheap one, circulation figures are dropping. This is mainly due to changes of interest toward daily politics, increasing inflation and a great increase in the number of newspapers and journals.

Part of the public which was sensitive to mathematical articles became involved in computers. For them various periodicals have been and are being established. There are weekly to quarterly periodicals. Most of them make irregular excursions into the world of algorithmic mathematics.

Editors of other periodicals frequently feel forced to publish mathematical writings. A political monthly, Valòsag (Reality, price 40 forints/number) published, e.g., "Dialogues about mathematics" by A. Rényi. In this paper the author explained his ideas and philosophy about the relation of mathematics to real life as a conversation between Socrates and Hippocrates. The chosen form and the target-oriented formulation resulted in a very good reception among people working in social sciences or simply in politics.

Two other dialogues written by A. Rényi about the applications of mathematics were published in the "Fizikai szemle" (Review of Physics, 15 forints per number). These were aimed at researchers in sciences, thus there was no need to follow the method of Socrates, but a constructed conversation between Galileo and Torricelli has attracted many readers.

7. ITEM 6: BOOKS

We discuss three types of books on popularization.

The first class addresses general questions like the basic problems and tasks of mathematics, possibilities for its applications, or the fundamentals of a given specific subject. The dialogues by Rényi belong to this kind of popularization. These works has been actually edited as a book, which has been translated into German, Russian and English, see [5]. The success of such books depends very much on their style. Specific expressions should be avoided or, at least, circumvented. The best is if they do not need great effort to read. This applies even if they have one specific theme. As a good example I would like to mention another work by Rényi: "Napló az információelmóletröl" (Diary on the Theory of Information, [6]). It also shows that such a book can (and should) discuss questions of daily politics. Another example is the Hungarian translation of "Lady Luck" by Warren Weaver [8].

Teasers appear to be the most attractive popular books. The fifth edition of "SICC" by József Grätzer [1] has appeared, e.g., in 108.000 copies in 1986 and it is already sold out. 2 - 4 such teasers appear in Hungary in a year, including translations. The quotation in the introduction was taken from such a book [3]. (The size of an edition is 17.000 copies, price in 1982: 16 forint, reduced to half price in 1986.)

The third type of popular book on mathematics comprises case-studies of applications. "Statistics by Examples", see [10], furnishes a good instance of this possibility. There are, however, just a few such Hungarian publications and even they are not very popular. The book "Tiz példa a matematika alkalmazásaira (Ten Examples for the Application of Mathematics) was hardly sold beyond the pseudo-obligatory purchase of libraries.

8. ITEM 7: FREE UNIVERSITY

There is a **numerus clausus** in the Hungarian universities, i.e. the number of students in all branches of higher education is bounded by government regulations. Therefore some kind of self-education should be offered to those who can not get it within the official system. This need has moved the TIT to establish an unofficial but free further educational institution, which has been named "József Attila Szabadegyetem" (Free University Attila József). This institution is free since it is open to everybody. It has just one defect: it is not a university by any means of the word. It can not issue acceptable certifications, there is no way of graduating in any subject. All the same it was popular for a long time, because it was the almost unique forum to get scientific knowledge.

The main activities of this free university are accomplished through courses organized within specific themes, and comprise 6,8,12 or 16 occasions. Participation in the mathematical courses is decreasing. In the capital with over 2 million population, a maximum of two courses can be organized. In most cities the society is turning over to a system of individual invited lectures, which are well-attended.

The activity of TIT is characterized by providing details on the distribution of lectures according to their subjects during the years 1981 - 85 including invited talks, clubs of young mathematicians and lectures at the free university:

Years	1981	1982	1983	1984	1985
Basic mathematics	70	53	76	129	61
Analysis	10	11	18	67	27
Algebra and Number theory	31	23	46	107	117
Probability and Statistics	17	32	40	42	9
Geometry	27	15	31	3	97
Topology	37	8	15	9	1
Set theory and Logic	40	144	128	81	63
Combinatorics	60	107	109	70	16
Applications case studies	6	16	17	1	2
History of mathematics	3	23	5	0	0
Didactics	33	7	0	10	4
Unclassified	13	26	45	35	20
Computers	72	107	247	1205	2581
Total	419	574	777	1787	3098
Number of participants	11027	15913	20579	33743	67353

9. ITEM 8: SEIZING THE OPPORTUNITY

As was mentioned in the introduction, a very important dimension of the popularizational work is the timing. We offer just two examples in this respect.

The first is concerned with computing percentages. This was a neglected area of school mathematics even before the "New Maths", and every-day practice hardly needed it. Slowly people were geting unfamiliar with it. Last year the changing economics policy has changed the one bank, one interest policy and now banks are offering a variety of interest rates. In addition to that, a new income tax system was introduced, forcing many people to calculate percentages. This situation created demand for lectures on this old and simple topic, which did not appear to be a serious topic for TIT lectures before.

This year's political reforms made it possible for some people to initiate the withdrawal of their delegate to the parliament. For this reason they had to collect 2.000 signatures and they actually collected more than 3.000. The administration wanted to check these signatures for correctness one by one. This was an excellent occasion to explain the idea of representative sampling in the daily newspapers. Since the administration was strongly against sample based decisions, the debate went on for a long period, and even members of the Academy of Science had to declare the power of representative sampling. Many people were converted to believe in the role and reliability of statistical methods.

References:

[1] Grätzer J. (1986) SICC, Móra Kiadó, Budapest.

[2] Howson A G, Kahane J-P, Pollak H. (1988) Popularization of Mathematics, Discussion Document, *L'Enseignement Mathématique*, **34**

[3] Ignatyev J I. (1982) A találékonyság birodalmában (In the empire of inventiveness), Tankönyvkiadó, Budapest.

[4] Nemetz T. (1988) Report on two Hungarian Contest for School Children (Age 10-14), *Mathematics Competitions*, Vol. 1, No. 2, 9-12.

[5] Rényi A. (1967) Dialogues on mathematics, Holden Day, San Francisco.

[6] Rényi A. (1984) A diary on information theory, Akadémiai Kiadó, Budapest.

[7] Swetz F J. (1978) Socialist Mathematics Education, Burgundy Press, Southampton (USA).

[8] Weaver W. (1979) Szernccse kisasszony, Gondolat, Budapest, (Lady Luck, Doubleyday and Company, New York, 1963)

[9] Csiszár I. et al (1967) Tiz példa a matematika alkalmazására Gondolat, Budapest.

[10] Mosteller F. (editor) (1967) Statistics by Example, Adison Wesley, Reading.

Sowing Mathematical Seeds in the Local Professional Community

A.G. SHANNON

University of Technology, Sydney, Australia

1. INTRODUCTION

Staff at this institution try to popularise mathematics in a number of ways. One is by visits to high schools, particularly those in rural areas, where talks are given on careers in mathematics other than teaching, and students are involved in a problem solving exercise in the style of Polya's film on guessing. Another method is Open Days and Information Evenings where the public can view and touch and have explained some of the research and consulting projects of the staff.

Another approach evolved when the writer was invited to speak at a local Rotary meeting as a last-minute replacement for a celebrity! The audience was bound to be disappointed with the absence of a real "personality", and even more with his replacement by an academic. Australian businessmen have a sometimes justified mistrust of "egg-heads".

My brief, if I accepted was to speak for 20 to 30 minutes, to make it interesting and to allow for some questions at the end. What could I talk about to a mixed well-fed audience, many of whom probably had unhappy memories of their own school mathematics and uneasy bewilderment of what their children were learning?

My professional work at the time ranged from teaching a class of first year students repeating a biometrics course to struggling with a nonhomogeneous, nonlinear recurrence relation with a research student. There was enough gloom around without using these as a starting point.

We are also encouraged to do industrial consulting. This helps our public profile (when successful) and brings in money to the university (and the individual). A few aspects of this seemed to have the right balance of optimism and interest. What had to be done firstly was to show in a few sentences that there was a problem, and then to outline how mathematics could contribute to a realistic solution.

Clearly the choice of speaker and topics is important. The latter should be something to which the audience can relate in some way, preferably with something tangible to show them, and the speaker should be able to address them with clarity, knowledge, enthusiasm and a touch of humour. The following are some topics which the present writer has used, though they are outlined here for a mathematical audience rather than a lay audience for whom lots of diagrams would be necessary, as well as working examples of the devices in question.

While there is always the danger of attempting too much in a short time, these audiences do not want to be talked down to, and once they have the problem explained they are happy to see the end result without too much detail in between. A useful by-product, though not actively sought, has been commercial help for some of our research and development activities. This, and invitations to talk at other venues and enquiries about buying the products, are tangible indications of success in this approach to the popularisation of mathematics amongst groups of business and community leaders who are, in Australia, increasingly becoming involved on committees which advise governments on educational policy issues.

2. THREE EXAMPLES

2.1 X-ray Camera

A Government department wanted a Gandolfi X-ray camera sample holder to oscillate through 90° in a vertical plane while it rotated through 360° in the horizontal plane to avoid coincidences of the crystal planes.

The solution was to utilise a property of the hypocycloid which is the locus of a fixed point on the circumference of a circle rolling (without slipping) on the circumference of a larger circle. When the radii are in the ratio 2:1, this locus is a diameter of the larger circle. The first condition could then be satisfied by a hypocycloidal gear mechanism in which a rod on the smaller circle was attached to a mechanism which could oscillate while the larger circle could rotate.

If the respective angular velocities are $\dot{\phi}$ and $\dot{\theta}$, then while a point on the larger circle does one revolution in $2\pi\dot{\theta}$ seconds, the rod takes $2\pi/\dot{\phi}$ seconds to traverse the groove. Coincidences occur when integer multiples of the two times are equal: $m\dot{\theta} = n\dot{\phi}$. By choosing relatively prime numbers for these integers, we can satisfy the second condition. (Two versions of the camera have been patented).

2.2 Credit Creation in NBFIs

The process of credit creation in the banking system is well known to economists. The conditions under which the deposit taking non-bank financial intermediaries (NBFIs), such as building societies, can create credit are less well known.

We shall assume that banks observe a constant liquid assets to deposits ratio (r) and that this ratio is always attainable. The requisite liquidity ratio is governed by legislation or convention or commercial prudence. Also assume that the banking system will gain s of any potential deposit while the NBFIs gain $(1 - s)$ of the deposit, where $0 < s < 1$. Thus, any deposits, whether from the banking system or the NBFIs, will be distributed between the banks and the NBFIs in the proportions s and $1 - s$. Assume again an extraneous injection of liquids into the banking system of D, and that the required liquidity ratio of the banking system is r_2 while that of the NBFIs is r_1, where $0 < r_2 < 1$ and $0 < r_1 < 1$. A bank approached us to determine the ultimate increase in deposits in the banking sector from this initial deposit D.

If a deposit D is made to a bank at stage 1, the required liquids will be Dr_2. If the bank is prepared to grant overdrafts to customers to the full extent of its liquidity ratio, then it can make an advance of $D(1 - r_2)$ which we suppose is redeposited. At the beginning of stage 2, $D(1 - r_2)s$ will be deposited in the banking system and $D(1 - r_2)(1 - s)$ in the NBFIs. The process quickly becomes unwieldy unless we recognise a pattern. The key is to use suitable notation and set up appropriate relations. If this is done, we can prove that the required answer is $D + Ds(1 - r_2)/(r_1 - s(r_1 - r_2))$.

2.3 Insulin Dosage Meter

The full extent of diabetes in Australia is not known. Mass screenings have been going on at shopping centres in recent years with mixed success. What is known is that the cost to Australia of diabetes is around 1.2 billion dollars annually.

According to a study at Sydney University Medical School "few diabetics were able to adjust their insulin dosage to overcome problems caused by intercurrent illness or differing dietary and exercise patterns. These problems in insulin administration present a serious barrier to a flexible lifestyle for patients with diabetes".

A drug company invited us to develop a computerised meter which would estimate a diabetic's daily short-acting and intermediate-acting insulin requirements based on information about blood sugar levels, exercise and carbohydrate variations and state of health. The devices we have developed do this and also modify the base doses of insulin. They also lock and store the information so that the physician can print it out later.

To do this, we initially set up a two compartment model for insulin-glucose dynamics in *vivo*. Later, other compartments had to be added to match experimental data. These enabled us to learn more about when insulins peaked in the body according to the injection site, and to look at the effect of varying mass indices.

The device we have produced is essentially a bar-code reader for scanning foods and their quantities for dietary analysis. This has a cap which can be connected to a commercial reflectance glucometer for blood glucose readings and to a modem for connection with a hospital computer.

3. DISCUSSION

The examples chosen can strike chords with most audiences. The X-ray camera is particularly useful in analysing metal fibre and the uses of this can be mentioned. Money matters make most of them sit up and listen, and many know someone who is a diabetic, and mostly there is at least one diabetic in the audience. Examples from statistics and operations research can also illustrate how powerful the mathematical sciences can be in sorting out the apparently imponderable.

There is usually a discussion during which someone nearly always says "I never knew what mathematicians did, other than teach mathematics". If there are tangible results, especially gadgets, for people to handle informally later this also helps. A simple poster on each topic can then be displayed too, provided it is professionally prepared.

It is important to bear in mind in such talks that one is not trying to teach mathematics. Rather, we show that there are problems to which mathematics can make a significant contribution. We may know this, but even a member of the Board of the Commonwealth Scientific and Industrial Research Organisation recently expressed doubts that mathematics had made any significant contribution to humanity in this century!

The approach has been refined as further invitations have come from different service clubs such as Apex and Lions. Since the members attend their meetings for reasons other than listening to the guest speaker, they are not necessarily put off by a topic involving mathematics, and one has an audience of influential local community and business leaders, hard-headed, but receptive to enthusiasm and clarity.

Mathematical News that's Fit to Print

LYNN ARTHUR STEEN

St. Olaf College, Northfield, MN55057, USA.

The theme of this session – understanding new trends, new results - invites reflection on two small Anglo-Saxon words: "new" and "news." "New" has several distinct meanings: not existing before; not known before; fresh; different; not old; of recent origin. "News" refers to tidings - to information about recent events.

The two words, in our context, reflect two professions: mathematician and journalist. Mathematicians deal with the new; journalists with news. Despite the common etymology of these words, in practice they have almost opposite meanings to the mathematician and the journalist. To understand new trends and new results, we have to examine how mathematicians and journalists differ in their perceptions of what's new and of what's news.

1. NEW MATHEMATICS

What is new to the mathematician? For some it is theorems - proofs of old conjectures or discoveries of new results. In 1983 it was Gerd Faltings' proof of the Mordell conjecture; in 1985 it was Louis De Branges' proof of the Bieberbach conjecture. In 1988 it was, for a short time, Yoichi Miyaoka's claim that he had proved Fermat's Last Theorem.

For others, what is new in mathematics are trends in research. For a good part of the 1970's, catastrophe theory was new; now attention has shifted to fractals and chaos. Forty years ago, many new trends in mathematics were expressed in the collective work of Nicholas Bourbaki as the culmination of

David Hilbert's agenda to provide a complete logical portrait of known mathematical theory. In the past quarter century, mathematics shifted once again to a counterpoint with applications - to nonlinear analysis and computational geometry, to spatial statistics and cryptography. Applications have spread from biology to finance, from fluids (flames, fusion, tornadoes) to data (stock markets, satellite transmissions, geological sensors).

Still for others, the frontier of mathematics is defined by new concepts or by syntheses of old concepts into significant new perspectives. Although algorithmic notions are not new, in 1971 when Stephen Cook wrote his seminal paper on complexity theory, the concept of NP-complete was brand new to mathematics. Although neither iteration nor dynamical systems are new - the former having deep roots in Newton's method, the latter in the work of Poincaré - the derivative concept of deterministic chaos is essentially new and potentially revolutionary. Although modelling and simulation are not new, the idea of computational science as a third paradigm paralleling experimental and theoretical methodology is new to the world of mathematics and science.

2. MATHEMATICAL NEWS

The journalist, in contrast, deals with news, which in this commercial age is not just a record - as the *New York Times* says - of all that's fit to print, but really a record of all that the public (or advertisers) are willing to purchase. What gets printed, by and large, is what interests the public.

There are, of course, many different publics reached by as many different media. What is of interest to one may be almost irrelevant to another. Magazines, newspapers, museums, radio, television, film, and books are all aspects of a vast and diverse media world. As mathematicians differ in their interpretation of the new, so journalists will differ in their view of news.

Serious intellectual magazines such as *Science* or *Nature* and newspapers such as the *New York Times*, *Le Mond*, or the *Guardian* do cover advances in mathematics - not often or regularly, but enough to meet a minimal professional obligation in which their editors believe. That they don't provide more coverage is largely due to limitations on audience interest, or to editor's perceptions of audience interest. Their readers want a potpourri of news, not too much of any one thing.

Most newspapers and magazines do not feel any obligation to cover mathematics (or even science) regularly. For them, mathematics is not a regular news beat the way health, politics and sports are. The difference is not because there is less news in mathematics, or less significant news, but because there is less interest among the readers. It is the readers, not the mathematicians or the journalists, who ultimately decide what news is fit to print.

Printed media do not, of course, reach the public at large. At best, they reach and influence public leaders. To reach a mass audience, one must use the media of the masses - primarily television. In a mass media the costs are higher and the common denominator of audience interest is lower. Together, these constraints virtually squeeze mathematics out of the picture. In the United States at least - where admittedly the common public intellectual level is not as high as in some other countries - I cannot ever recall seeing strictly mathematical news on either a commercial or public television news show.

I can summarize these reflections by a sweeping generalization: what the mathematician considers new, the journalist does not consider fit to print. There are exceptions, of course, but by any reasonable standard of measurement applied to the planet's five billion inhabitants, these exceptions amount to a set of measure zero, or at most epsilon.

3. CASE STUDY

Rather than continuing to dwell in generalities and abstractions, I think it might be helpful to examine these issues in terms of a case study with which

I am familiar - the effort by the mathematical community in the United States to increase media coverage of mathematics. Our experience, with its share of successes and failures, illustrates most of the concerns that would face any country embarking on a similar venture. It also highlights intellectually interesting issues of what should be covered, what actually is covered, and how that coverage is interpreted.

Before beginning I should say that this recital of events is not by any respectable standard an objective report. I have been in the middle of many of these events, sometimes as a quasi-journalist, sometimes as an advisor, and occasionally as the responsible decision-maker. One of the responsibilities of an alert audience is to detect bias and filter it out.

I'll begin the story at the point where I know it best - in the early 1970's - although the real roots go back to events following the end of the Second World War. In the 1970's, Saunders Mac Lane was very active in political leadership of the American mathematical community. He was one of the very few persons who had been president of both the American Mathematical Society and the Mathematical Association of America, and he was a member of the governing board of the U.S. National Science Foundation, the government agency that awards grants for research and education in science, mathematics, and engineering. For a period of time he also served as Vice President of the National Academy of Sciences, a private self-governing organization chartered by the Congress of the United States during the presidential term of Abraham Lincoln to provide independent scientific advice to the Congress and people of the United States.

Mac Lane became convinced that one reason the U.S. mathematical community was having difficulty in securing appropriate financial support for research and education was that those who set national policy knew virtually nothing about the nature of mathematics nor its benefits to society. At his instigation, the Conference Board of the Mathematical Sciences (CBMS) - a consortium of a dozen or so different mathematical professional societies - successfully sought support from the National Science Foundation for a project to explore and promote public understanding of mathematics.

At that time, about 1974, there was virtually no coverage of mathematics in newspapers in the United States, nor was there much coverage in general scientific periodicals. Allen Hammond, who had just finished a Ph.D. in applied mathematics (geophysics) at Harvard, had been hired by *Science* magazine a few years earlier to edit the Research News section; he hired Gina Bari Kolata who had master's degrees in biology and in mathematics (and who was the wife and daughter of mathematicians). Between them they produced a few articles a year about mathematics - and that was essentially all there was.

I worked part-time for CBMS for a few years on the NSF project, making contacts and testing the water. We held a meeting at the Ontario Science Center for leaders of science museums; we arranged for a seminar on mathematics at the annual meeting of the Council for the Advancement of Science Writing, a small professional society of science journalists; we discussed ideas for mathematical topics with the producers of NOVA, a television science series produced in England by BBC and in the United States by WGBH in Boston; we fed lots of story ideas and actual draft articles to editors and science writers at the *New York Times, Scientific American, Science News,* and *Science*; and we produced a prototype mathematical magazine called *Mathematical World* that helped pave the way for Springer-Verlag's *The Mathematical Intelligencer*.

4. CATASTROPHES

One anecdote from this period illustrates the problems we faced. *Science News* is a small but important publication in the United States that produces each week about sixteen pages of research news written in a style that makes it suitable for use by high school and college students and teachers. It has a large circulation among science teachers in school, so is more influential than the better known AAAS journal *Science* in attracting students to careers in science.

When I was introduced to Kendrick Frazier, editor of *Science News*, he explained that he did not cover mathematics because it was impossible: there were too few stories of interest to their readers, and no one could make even these few stories clear to their audience of students and teachers. He did agree, however, to read some samples that I promised to send as a trial to see if perhaps it could be done.

The first thing I sent him, in September of 1974, was a very short news report from the Vancouver meeting of the International Congress of Mathematicians based on Christopher Zeeman's talk on catastrophe theory. That, I thought, had sufficient appeal to interest even the most jaded editor.

Two things happened in the months that followed. First, *Science News* printed the report and, several months later, a follow-up. It turned out that they received more mail on that story than on any other they ran that year. I received letters from all over the United States, and all over the world, especially from scientists and teachers who wanted additional references and further information. As an existence proof that mathematics can be made interesting, it succeeded.

But then there were the mathematicians. Hirsh Cohen, I remember, expressed concern that this wasn't new mathematics. It was simply a rediscovery or repackaging of some very old ideas about dynamical systems due to Poincaré. There was some evidence that even Laplace and Euler knew of these phenomena. So it was a gross disservice to mathematics to single out catastrophe theory as one of the very few examples that the public would see that year about contemporary mathematical research. The news wasn't really new.

Other mathematicians were not so polite. They excoriated the purveyors and reporters of catastrophe theory for spreading mathematical malpractice by suggesting applications, especially in the social sciences, that could not be sustained by scientific theory. In the physical sciences where there is a potential function that ensures that the mathematical theory paralleled a scientific model, they argued that the catastrophe theory is legitimate but old; in the social and behavioral sciences where there is no obvious potential function, they argued that catastrophe theory is pseudo-science. In either case, the news about catastrophe theory was not fit to print.

The public, of course, did not understand what the argument was all about. They did hear from many sources about this latest trend in mathematics and about some of its more exotic applications. For many, it was their first realization that mathematicians actually do something other than teach. The public began to realize that mathematics, like biology and physics, is an active area of research and that mathematical research, like research in biology and physics, is beyond their comprehension. DNA, quarks, and catastrophes were comparable abstractions in the public mind. They didn't understand any of it, but they did recognize that it represented the frontier of science.

From my point of view, the catastrophe stories were far from a catastrophe for mathematics. Indeed, they served their central purpose very well: to awake a sleeping public to the fact that mathematics is an area of research just as active and potentially just as interesting as any other.

5. DISTORTIONS

The CBMS project eventually produced the volume *Mathematics Today*, which led to its sequel *Mathematics Tomorrow*. My forays into mathematical journalism succeeded well enough that I served as mathematics correspondent of *Science News* for a period of about six years, writing occasional pieces as I found time and opportunity. Fortunately, Ivars Peterson joined *Science News* as a summer intern just about the time that I was getting too busy with other matters to sustain this work. He had a strong background in physics and mathematics and was eager to take over this beat, which he has done with distinction. Some of you may have seen his recent book *The Mathematical Tourist*, which retells many of the stories that he has written for *Science News* throughout the 1980's.

Before returning to the chronology of events in the United States that has literally transformed the level of press coverage of mathematics, I'd like to discuss a second anecdote that reveals yet another pitfall - distortion. In 1979 I received a tip from someone, probably Ron Graham, about the publication by the Russian mathematician L.G. Khachian of what has since come to be known as the ellipsoid algorithm for solving linear programming problems. On October 6 I wrote an article on this surprising discovery for *Science News*; a month later, Gina Kolata wrote one for *Science*, and shortly thereafter Malcolm Browne wrote a front-page story for *The New York Times*. Each of us engaged in slight journalistic oversights in the interests of clarity - since surely none of the readers could absorb a full mathematical statement of the result. But as the story progressed from mathematically-trained writers to generalists, the border between innocuous simplification and dangerous distortion was crossed by writers and editors who did not know enough mathematics to understand the story.

One can follow the progression of exaggeration in the three headlines: In *Science News*, my story appeared under "Linear Programming: Solid New Algorithm;" in *Science* it became "Mathematicians Amazed by Russian's Discovery;" while the *Times* proclaimed across four columns of the front page, "Soviet Discovery Rocks World of Mathematics." By the time the story got to the *Times* - and also to the *Guardian* in England - it sounded as if the Russians had discovered a secret for solving the travelling salesman problem and cracking secret codes. "This fact has obvious importance for intelligence agencies everywhere," reported the *Times* in ominous language.

This story has certain subtle features. First, in 1979, it was rather well known - even among science journalists - that some important problems were intractable even for the fastest computers. Second, linear programming was a tool of enormous economic significance in the oil industry, in transportation, and in defense. Third, the protagonist in this story was Russian, and the media were in the United States. The story broke just as Ronald Reagan began his campaign for President under an anti-Soviet "evil empire" theme.

The correct mathematical news in this story was that no one knew, prior to Khachian's discovery, whether linear programming did or did not belong to the class of NP-complete problems - the intractable ones. There was no known polynomial-time algorithm, but no proof that none existed. We did know that the simplex method, the mainstay of effective LP algorithms, was not in itself polynomial time for all possible inputs: it was easy to construct examples, albeit quite artificial, for which the simplex method took exponential time to converge.

The ellipsoid (or interior) algorithm was the first algorithm guaranteed to converge in polynomial time for all LP problems, both realistic and artificial. So the mathematical news in this story was a proof that LP was in the class of polynomial time problems - those for which a polynomial time algorithm is known to exist. The new method held promise of improving on the simplex method in some cases, especially in the integer-programming variants which are intrinsically more difficult, but efficiency depended critically on concrete computational details since the interior methods involved a lot of matrix inversions. So actual performance would depend greatly both on the particular problem and on the efficiency of the particular computer code. (As it turned out, some years later Narendra Karmarkar at Bell Labs discovered a new approach using the distortions of projective geometry to produce a truly efficient interior algorithm for linear programming. But this gets ahead of the story as it unfolded in 1979.)

The distinction between the general theorem (for the first time, LP was known to be not intractable) and the particular case (although polynomial-time, the ellipsoid algorithm might compute more slowly than the exponential-time simplex method because of details of input data and coding algorithms) was generally too fine for science journalists (or their readers) to catch. My story was accurate, albeit just barely. Berkeley computer scientist Eugene Lawler, in an analysis of this episode in *The Sciences* (September 1980) said of my report that it was "generally correct, and did not seriously mis-state the significance of the achievement." Gina Kolata's report had one unfortunate sentence claiming that Khachian's result is "tied to" the infamous travelling salesman problem (which, of course, is tied to encryption algorithms). Malcolm Browne in the *Times* - the third in the series - interpreted that link as a solution, and then set off speculation about the Russians beating the U.S. to a key computer code for industrial competitiveness.

I need not describe the backlash that these stories caused in the U.S. mathematical community. After being inundated with letters and briefings by mathematicians, the *Times* eventually printed what it considered to be a correction, saying that further analysis by American mathematicians revealed the result to be "'far from the seminal achievement originally portrayed." They coyly refrained from mentioning who it was - mathematician or journalist − who painted the original portrait.

Despite the errors and distortions, I would not be the first to criticize Browne or the *Times*. Ultimately, the public doesn't remember the details, whether right or wrong. They *do* remember, however, that mathematics has something to do with industrial efficiency, that computer codes are things that mathematicians work on, and that competition in these areas is an important part of the East-West political game. So again, the long-term goal of portraying mathematics as an active, interesting field of significance to society is achieved. Who cares if the algorithm is as bad as n^6 ?

6. COMMITMENT TO ACTION

Many of the stories that found their way into the U.S. press during the 1970's and early 1980's were due to tips from one person: Ronald Graham. Graham went to a lot of meetings, he knew the few journalists and mathematicians who were writing general stories, and he believed in the importance of this activity. Without his consistent tips, the ϵ coverage of this period would have been only $\epsilon/5$.

Of course the price of tips from Ron Graham was a view of mathematics from the perspective of discrete mathematics. Algorithms and number theory thrived in the press; non-linear analysis and geometry did not do as well. But again, that's not so bad, since most of what emerged in the news was not just a sample of new mathematics, but a sample drawn from new areas of mathematics. So the alert public who followed these stories - mostly scientists, by the way - got the message that mathematics was not only thriving, but expanding into areas untouched by their school experience. That's not a bad message to receive, however slighted it may make the PDE folks feel.

It soon became clear to leaders of the mathematical community that they could not rely on the press to provide the level of coverage needed to turn around what was becoming a critical situation in the United States for mathematics research and mathematics education. The press does not serve any external community. In theory, it serves the public interest; often it serves only its own interests. But never does it serve mathematicians' interests. If mathematicians wanted more coverage, they would have to do something about it by themselves.

The Joint Policy Board for Mathematics - a joint action committee of the three major university-level mathematical societies in the United States (AMS, MAA, SIAM) - established a small committee consisting of Ron Graham, Joe Keller and me to make recommendations. Our chief recommendation was that the mathematical community needed to do what every other scientific community had done: establish an office of external relations that included a professional who knew how to deal with the media.

There were two thrusts to this message. First, the mathematical community must do this together, since it would make no sense to splinter efforts among various groups, each with its own private agenda. Second, the lead person should not be a mathematician, but a professional experienced with print and video media. Mathematicians have little expertise in promoting stories to the media, and the few examples we had showed that they were incapable of learning.

I cite as an example the American Mathematical Society which, year in and year out, sent press releases to journalists about major speakers at its national meetings. These communications, in fine eight-point single spaced print (rather than editable twelve-point double spaced type) consisted of a concise abstract of the talk, a list of the speaker's previous publications, and a mathematical biography - all in language that only a specialist in the speaker's field could understand.

It is not surprising that no journalists every chose to come to AMS meetings on the basis of these types of releases. What might be a surprise is that no one with authority in the Society recognized the futility of this approach. Many observers of mathematicians claim that even this is not really a surprise - that mathematicians as a group are constitutionally indisposed to understand what motivates ordinary people.

The recommendation of our Committee was accepted by the societies, and gradually implemented. Kathleen Holmay, a public relations consultant who specializes in science issues, was hired in 1985 under part-time contract for the Joint Policy Board for Mathematics. It was a fortuitous time, for it came right on the heels of two major documents that focussed U.S. attention on mathematics and mathematics education.

In 1983 a report *A Nation at Risk* awoke the American public to serious and seemingly irreversible problems in our educational system. In 1984 the National Academy of Sciences released *Renewing U.S. Mathematics: Critical Resource for the Future*, what we all call the "David Report" after its committee chairman Edward E. David, former Science Advisor to President Nixon. The first report called for major overhaul of the nation's educational system, the second for major increase in support for mathematical research.

These two themes provided the spark needed to mount an effective public information campaign. They gave Kathleen Holmay an agenda that resonated with interests of the public: mathematics education, and international competitiveness rooted in mathematical sciences. By persistent and clever campaigns, she has managed to entice dozens of reporters to take on mathematics as one of their beats. In the last five years, press coverage of mathematics and mathematics education in the United States has increased by at least an order of magnitude.

7. CONSEQUENCES

Let no one think that this increase was due to a compelling public or reportorial interest in mathematics. Reader interest still runs to issues that affect lives such as cancer, global warming, or AIDS; journalists still view mathematicians with suspicion and latent hostility from their own school and college experiences. That's a risk we will always run: everyone who is not a mathematician probably stopped studying mathematics as the result of a particularly unpleasant experience in school. No other subject has such a legacy of negativism to overcome.

What success U.S. mathematicians have had in publicizing mathematics is the result of two things: a competent professional in place to make the news flow, and an orchestrated climate of crisis to make the media receptive to the news. Reporters will not be able to convince their editors to make space for mathematics in competition with news about potential cures for AIDS unless we give them substantial, documented reasons why mathematics is just as important.

We have a saying that expresses well just what's going on - at least if we read it backwards. The saying is: "No news is good news." In mathematical journalism, the reverse is true: "Good news is no news."

Some of the crisis talk is hype, but much is not. U.S. standings in international comparisons of mathematics education are dismal. Our ranking among nations is like our balance of payments: below all our competitors, and just barely above some third world countries. During the 1970's, the number of college graduates with mathematics majors fell by over 50%, as did the number of U.S. students who continue on to a Ph.D. in mathematics. The U.S. National Security Agency - the formerly super-secret enterprise that deals with encryption of military secrets - has been trying to hire almost as many mathematicians as we produce each year, but without success.

The good side of these dreary figures is that it is driving salaries for mathematicians up, at least in institutions with sufficient resources to compete for the best people. And it is opening the pages of the popular press to stories about mathematics. The wedge that creates the opening is mathematics education, since everyone with children in school has opinions about education. From there, reporters can move naturally into how computation is changing the nature of mathematics, and then into stories about news in mathematics.

Overall there has been an increase from 2-3 to 16-20 in the number of reporters in the United States who take seriously news in mathematics, and an increase from under twenty to several hundred in the number of stories about mathematics and mathematics education that appear annually in magazines and newspapers addressed to general audiences. Much of this success has been built on a series of events created or orchestrated by the mathematical community:

- 1986 The Congress of the United States and President Reagan declare Mathematics Awareness Week in the month of April.

- 1988 The Centennial of the American Mathematical Society provides an excuse for a year-long, one-event-per-month focus on mathematics research.

1989 Publication of *Everybody Counts* and *Curriculum and Evaluation Standards for School Mathematics* provides a reason to focus on issues in mathematics education.

8. CONTROVERSY

As one might expect, publicity about mathematics has not been achieved without controversy. The lines that Ms. Holmay casts to journalists new to this field sometimes snare mathematicians who recoil in disgust at the worm they find on the end of the hook.

At the International Congress of Mathematicians in Berkeley in 1986, a press release opened with appealing comments about numbers, statistics and the consumer price index as an inducement to attend a special pre-Congress talk on modular forms. For the Centennial of the American Mathematical Society, which opened on the eighth day of August, the eighth month, in 1988, the press office put out a clever piece by Martin Gardner on "Dr. Matrix and the Wonders of 8," a spoof on numerology that many reporters thought was serious. But they bought it, as we say, "hook, line and sinker." On the local television news there was a feature story whose lead was, as one might expect, "8/8/88 - the things mathematicians do for a living." Without such a lead, mathematics may never have made the news at all.

Ms. Holmay works directly under Kenneth Hoffman, who served for five years as Director of the Office of Government and Public Affairs of the Joint Policy Board. Together they orchestrate the now-annual Mathematics Awareness week using press releases, posters, postcards, school activities, and whatever promotional events they can dream up. Each mathematics awareness week has a special theme, backed up by a poster that teachers can hang on bulletin boards. This year's theme was "Discovering Patterns;" it features information from Branko Grunbaum based on his book *Tilings and Patterns*, an article that I wrote for *Science* entitled "The Science of Patterns," and special material written by Ian Stewart on "The Impact of Mathematics." (Despite 200 years of independence, Americans must still turn to England for an occasional infusion of our mother tongue.)

The effort of the last decade is paying off. Hoffman and Holmay serve as a semi-permeable membrane separating the arcane world of mathematics from the homely world of journalism. They develop themes and invent story lines that resonate with both public interest and mathematical events; they introduce journalists to mathematicians and mathematics educators, and provide essential background information to journalists who know nothing about the world of mathematics. They also fend off attacks from mathematicians insensed over numerology, and parry thrusts by journalists who want to limit school mathematics to consumer topics. Within the world of mathematics, they make news of what's new.

9. LESSONS

One can read these events in many ways. By the standards of other sciences, popularization of mathematics is still an insignificant fraction of total science journalism in the United States. But by the standards of the early 1970's, today's public knows vastly more about the importance of mathematics in school and the role played by mathematics in society. So we have made much progress, yet there is still a long way to go. As we move along this largely unblazed trail of popularization of mathematics, I commend to you several lessons from our experience in the United States:

1. *Mathematicians make lousy publicists.* If you want the job done, hire a professional. Mathematicians who serve as publicists are always shadowed by their colleagues' standards. Despite inevitable distortion, non-mathematicians will almost always do a better job of actually communicating with the public.

2. *Theorems won't sell in a vacuum.* If you want to interest the public in real mathematics, first get their attention with something closer to their heart - like education, economy, or environment.

3. *Literal truth is irrelevant.* The purpose of popularization is to raise awareness, not to educate. What must be communicated is not the letter but the spirit of mathematics. The criteria of success is not an increase in knowledge, but a change in attitudes.

4. *Don't underestimate public interest in mathematics.* Everyone has studied some mathematics; many are amateur mathematicians, some even closet amateurs. There is more public interest in mathematics than editors realize or admit.

5. *Don't underestimate public ignorance of mathematics.* Most people don't even know that mathematics is a living discipline. Their image of the subject is locked in the age of Euclid or Newton, framed by school experience of set problems and mechanical worksheets. Changing this image is a sufficient and worthy goal of any program to popularize mathematics.

6. *Don't pander to utilitarianism.* Editors and reporters often judge news by immediate utility. While utility is a legitimate value of mathematics, immediate utility is not. Don't be drawn into dishonest claims of cures for cancer or economic miracles as the consequence of the latest breakthrough in mathematics.

7. *News need not be new.* Rarely do important trends become visible overnight. The impact of computing on statistics and on non-linear analysis has been gradual, not sudden, but is no less news for that reason.

8. *Connect with school mathematics.* School is part of everyone's experience, for good or ill, so it provides a common base of discourse. New mathematics inevitably suggests possible new ideas for school curricula, which can serve as a news peg for journalists.

9. *Highlight legitimate applications.* Applications appeal to multiple journalistic beats - to science or health or economics. Good mathematics shines through good applications.

10. *Stage news-worthy events.* Since reporters need an "event" to claim scarce space in the press of daily events, meetings should be planned to provide reporters with a legitimate news peg.

Christmas Lectures and Mathematics Masterclasses

E. C. ZEEMAN FRS

Hertford College, Oxford, U.K.

I was thrown in at the deep end of popularization by having to give the Royal Institution Christmas Lectures [5] in 1978. The Christmas Lectures were founded by Michael Faraday over 150 years ago in order to give schoolchildren an opportunity to see some science. They have been given every Christmas since then (except during the war). They are usually devoted to physics, chemistry, biology or medicine, and mine were the only time they have ever been given on mathematics. There are six one-hour lectures to an audience of Young Persons at the Royal Institution between the ages of 10 and 17, and nowadays they are also broadcast on TV each day between Christmas and New Year, or thereabouts. So it was a daunting task: they had to be accessible not only to schoolchildren but also to the general TV public, and subject to the direct scrutiny of one's colleagues and other professionals.

What kind of mathematics could one possible give on TV to such an audience? It took me a year to select the topics, and about six months to write the lectures, with a month of panic at the end preparing all the demonstrations and illustrations. Fortunately I had the time because I happened to be in the middle of a 5-year SERC Senior Research Fellowship. The Chairman of the SERC, however, wrote to me afterwards ticking me off for wasting my time popularizing on TV instead of doing research! I wrote an equally vigorous letter back saying that I thought it was very important to bring mathematics to the attention of the public, particularly to young persons, and if the SERC couldn't allow me 10% of my time to seize this unique opportunity then they were suffering from a regrettable lack of vision. Having to

Lectures and Masterclasses

write the lectures, I added, had in fact unexpectedly stimulated some original research in differential equations and population dynamics, as I shall explain below.

Returning to the business of choosing the topics I thought it would be appropriate to devote the first three lectures to pure mathematics, and the last three to applied. In selecting the pure material I ran into an endemic problem with the BBC: the argument lasted several months but (with apologies to my good friends at the BBC) I will paraphrase it in dialogue form.

Z: Mathematics is about theorems and proofs[*], so I should really give some famous proofs.
BBC: But we can't have you in front of a blackboard with the back of your head to the camera.
Z: That's OK, I'll use an overhead projector.
BBC: No, no, it'll be too much of a school image.
Z: But proofs are beautiful and can be deeply inspiring.
BBC: You miss the point: we have to *entertain* the audience otherwise they'll switch off.
Z: But proofs *can* be entertaining, and indeed enjoyable and riveting.
BBC: We strongly advise you not to give proofs.
Z: I insist on giving proofs.
BBC: We forbid you to give proofs.
Z: In that case you can go jump in the lake, because I'm paid by the RI to lecture to an audience of young persons, and if you don't want to broadcast it you needn't.

This is the only time I have ever won a serious argument with TV producers: they usually have the upper hand because they can subsequently edit out anything they don't like, but the Christmas Lectures are broadcast neat without editing out anything except perhaps the odd cough.

Now came the question of what theorems to choose? That caused me many sleepless nights. The danger of popularization is triviality, especially if too much

---

[*] As Douglas Quadling [2] says (page 2) any move to popularize mathematics should have as a major goal an appreciation of the central place of proof.

emphasis is given to games. As Michele Emmer [1] says (page 16) it is important not to give the impression that mathematics is merely a game. Eventually I worked out some criteria.

Criteria for choice of theorems in the Christmas Lectures
1. A theorem should be noble. In other words, it should capture the quintessence of some mainstream branch of mathematics.
2. Results should be surprising and intriguing, sufficient to capture the imagination and hold the attention.
3. Proofs should be elegant, rigorous, complete and understandable.
4. Each proof should fit onto a single transparency.

Having chosen the theorems, I then grouped them into three lectures, one on topology, one on numbers, and one on infinity.

Examples of some of the theorems
1. The existence and uniqueness of linking numbers (with application to circular DNA).
2(a). The irrationality of $\sqrt{2}$.
2(b). The algebraic closure of the complex numbers.
3(a). The divergence of $\sum \frac{1}{n}$.
3(b). Cantor's theorem for infinite sets.

For example, let us look at Cantor's Theorem, and see how to fit the proof onto one transparency.

       Let A and B be sets, finite or infinite.

<u>Definitions</u>
$\quad\quad\quad A = B$ if there exists a bijective map $\ A \to B$.
$\quad\quad\quad A \leq B$ if there exists an injective map $\ A \to B$.
$\quad\quad\quad A < B$ if $\ A \leq B\ $ and $\ A \neq B$.
$\quad\quad\quad 2^A$ is the set of subsets of A.

<u>Theorem</u>$\quad A < 2^A$.

Proof $A \leq 2^A$ by the map $a \to \{a\}$.
Suppose $A = 2^A$ by a bijective map $f : A \to 2^A$.
Let $\quad B = \{a \in A \,;\, a \notin fa\}$.
Let $\quad fb = B$.

Then $\quad b \in B \Rightarrow b \in fb \Rightarrow b \notin B$

$\hfill$ contradiction.

$\qquad\quad b \notin B \Rightarrow b \notin fb \Rightarrow b \in B$

Therefore $\quad A \neq 2^A$.
Therefore $\quad A < 2^A$.

Before giving the above proof I used some volunteers from the audience to illustrate sets and subsets of boys and girls, and illustrated injective and bijective maps by matching them up. The proof for finite sets was illustrated by the barber's paradox: if the barber (b) shaves everyone (B) who doesn't shave himself, then who shaves the barber? ($b \in B$?) I took a set of three girls (A), and eight boys to represent their set of subsets (2^A); the girls chose partners (f), and we worked out who was the barber (B), gave him a white coat, and verified that the poor fellow had no partner (b).

When I came to the proof proper I wrote out the transparency line by line so that the proof unfolded before their eyes. The great danger of using an overhead projector with prepared transparencies is the tendency to bombard the audience with too much material all at once, causing psychological overload.

The beauty of Cantor's Theorem is that it immediately implies the astonishing fact that there are an infinite number of different sized infinities:
$$A < 2^A < 2^{(2^A)} < \ldots$$
Indeed this was so shocking to many mathematicians in the 1870's when Cantor first produced his theory of the infinite that he was violently and shamefully attacked, causing him to suffer a nervous breakdown.

Turning to the applied lectures I divided applied maths into four types, depending upon whether the application was concerned with discrete or continuous things, and whether they were behaving discretely or continuously. The figure shows a typical application in each box, and a typical branch of mathematics that would be used to model the phenomenon.

		THINGS	
		Discrete	Continuous
BEHAVIOUR	Discrete	Dice Probability Game theory	Music Discontinuities Harmonic analysis Quantum theory Catastrophe theory
	Continuous	Planets Populations Ordinary differential equations	Waves Partial differential equations

The most interesting box is the top right because it is not at all obvious why continuous things should behave discretely. Indeed nearly every branch of mathematics that has been developed to handle phenomena in this box has aroused some sort of controversy when it was first introduced. For example in the middle of the 18th century when Daniel Bernoulli first suggested that the shape of a violin string could be modelled by a trigonometric series giving rise to the harmonics Euler and D'Alembert protested vigorously that this was far too restrictive a notion compared with the generality of all possible functions. The battle then raged for fifty years until 1807 when Fourier wrote his thesis showing how to calculate the coefficients of the series, and eventually Dirichlet proved in 1822 that the resulting series converged to the original function, thus entirely validating Bernoulli's original point of view.

The introduction of quantum theory generated a similar controversy amongst physicists: it gave good predictions but the probabilities involved in the foundations were aesthetically so unattractive that Einstein refused to accept them all his life, saying that God doesn't play dice. More recently the applications of catastrophe theory to the biological and behavioural sciences have caused controversy amongst mathematicians, in spite of the fact that the underlying theorems are totally rigorous as well as being surprising, profound and beautiful.

In preparing the applied lectures I selected something from each box. The first lecture included examples of probability and game theory, and the game I chose was Maynard Smith's intriguing dove-hawk-bully-retaliator game which he invented to model the evolution of strategies in animal behaviour. It was when I was actually writing the lectures that I suddenly realized with amusement that this particular game was in the wrong box, because the proportions of the population playing each strategy were in fact changing with time, in other words they were *discrete* objects behaving *continuously* as they evolved towards an equilibrium. Therefore, by my philosophy, they ought to be in the bottom left box, modelled by an ordinary differential equation. So I immediately wrote down this equation and began to explore its properties; I found to my surprise that Maynard Smith's game was unstable and I showed how to stabilize it [7]. It transpired that I had discovered what are now called the *replicator* equations, and I managed to classify them in 2-dimensions [6]. At about the same time they were independently discovered by groups in Austria and America, who proved results complementary to my classification, and just recently my daughter Mary Lou Zeeman has extended the classification to 3-dimensions [10]. I mention this to illustrate the interrelationship between teaching and research: research often enhances teaching, and conversely teaching can often stimulate research.

Returning to the three Christmas Lectures on applied mathematics, the second one was about waves, harmonic analysis and music, and the third one was about my own work on the applications of catastrophe theory [4].

I had some difficulty in choosing a name for the series. I wanted to call them "The nature of mathematics and the mathematics of nature" but this was deemed to be too long. Eventually I settled on "Mathematics into pictures" because this emphasized the importance of the geometric approach and the powerful insight that it gives to every branch of mathematics; also I happen to love geometry.

Mathematics Masterclasses

Out of the Christmas Lectures grew the concept of the Royal Institution Mathematics Masterclasses. They were the brain-child of Sir George Porter, the then Director of the Royal Institution, who seized the opportunity to respond to the demand by young people for more mathematics that the Christmas Lectures had created. The classes at the RI have been going on ever since, and have now spread to more than twenty other places around the country. For instance we regularly ran classes when I was at the University of Warwick.

They are aimed at 13-year-olds because that is the age when abstract concepts can be mastered. The purpose is to provide enrichment for the more gifted, because today the most gifted can be amongst the educationally most deprived. The idea is to invite about 40 local schools each to pick their best 13-year-old mathematician, and these are invited to come for two and a half hours each Saturday morning for ten weeks. The style is to have talks interleaved with problem sessions, developing a single topic each Saturday, with different topics each week.

The most important part is of course the problem sessions, because that is when the young people have the opportunity to be creative and to discover things for themselves. I used to have half a dozen research students mingling with the young people during the problem sessions, and they liked that. One little girl came up to me and whispered "It's lovely to be amongst people who don't think you're odd". In a follow-up study four years later Watson [3] found that all the participants reported in retrospect that the masterclass experience had greatly increased both their confidence and their problem-solving ability in all branches of science.

Lectures and Masterclasses

Everyone who gives masterclasses has their own style and their own favourite topics. I myself like to weave mine around noble theorems, as in the Christmas Lectures. But of course one has the additional time and opportunity to stimulate discovery and creativity in the young people themselves, by means of the worksheets and problem sessions. I also like to do experiments. Let me summarise my own preferences, and describe some of my favourite classes.

Criteria for a masterclass
1. Noble theorems.
2. Results should be surprising and intriguing.
3. Proofs should be short, elegant, rigorous, complete, and understandable.
4. Worksheets should contain both theory and computation, and should stimulate discovery and creativity.
5. Applications should be useful.
6. Experiments should be surprising, robust, easy to do, and easy to repeat at home with home-made equipment.

Examples of masterclasses
1. <u>Perspective</u>. Proof of existence and uniqueness of vanishing points and observation points, and demonstrations of them.
2. <u>Gyroscopes</u>. Proof (without using calculus) of the gyro law from Newton's law of motion: that the spin axis chases the torque-axis. Prediction of precession times. Applications to tops, eggs and boomerangs.
3. <u>Gears</u>. Bicycle 3-speeds and car gearboxes. Planetary and differential gears (introducing simultaneous linear equations). The geometry of involute gears.
4. <u>Knots and links</u>. Proof of the existence of knots, and the calculation of knot numbers to distinguish between knots. The existence and uniqueness of linking numbers.
5. <u>Curves of constant diameter</u>. Proof that the perimeter is always π times the diameter. Proof that the circle is the curve with the largest area.
6. <u>Spherical geometry</u>.

As an illustration let me describe part of the last one on spherical geometry. I begin by proving that the angles in a triangle add up to 180°. They all know this already, and so they find it nice and reassuring, even a little boring. Then I surprise them with a spherical triangle containing three right-angles, whose angles therefore add up to 270°.

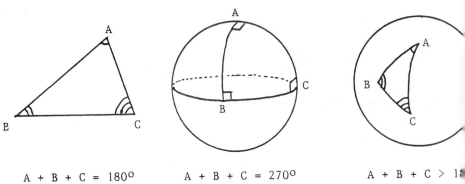

A + B + C = 180°   A + B + C = 270°   A + B + C > 18

Then I tell them there is a theorem that is true for any spherical triangle that expresses the sum of the angles as a function of the ratio of the area of the triangle to the area of the sphere, and I challenge them to find it in the next ten minutes. And to my astonishment they always do. There are always two or three in the class who discover the correct statement of the theorem, which is given below.

The interesting aspect of this experiment is that it plunges the class immediately into the typical predicament of a research mathematician. It is often the case that a familiar theorem no longer holds in a more general setting, and then one has to fish around and find the proper generalization for that setting. And the young people do just that: they experiment with a few examples, such as dividing the three-right-angled triangle in half, and half again, until they guess the pattern. Of course I do not expect them to discover the proof of the theorem, although there is a beautifully simple proof which I shall now give.

<u>Definition</u>. A *great-circle* on a sphere is the intersection of the sphere with a plane through its centre. A *spherical triangle* consists of the arcs of three great-circles. Let A, B, C denote the angles of the spherical triangle, T its area, and S the area of the sphere.

<u>Theorem</u> $A + B + C = (1 + 4(T/S))180°$.

Lectures and Masterclasses 203

Proof. Define the *A-lune* to be the area between the two great-circles through the vertex A, shown shaded in both the diagrams below.

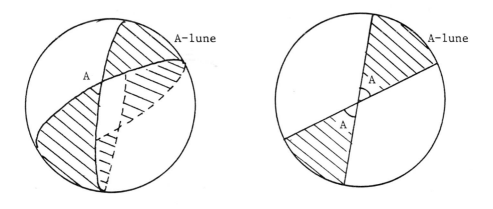

From the right hand picture one can see that $\dfrac{\text{A-lune}}{S} = \dfrac{2A}{360} = \dfrac{A}{180}$.

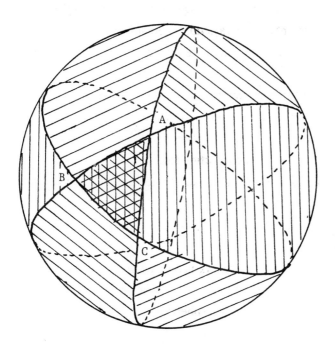

From the above picture it can be seen that the union of the three lunes covers the sphere, but covers the triangle three times, which is two times too many, and the antipodal triangle similarly.

Therefore $\qquad$ A-lune + B-lune + C-lune = $S + 4T$.

Dividing by S gives $\quad \dfrac{\text{A-lune}}{S} + \dfrac{\text{B-lune}}{S} + \dfrac{\text{C-lune}}{S} = 1 + 4(T/S)$

Substituting the formula above:

$$\dfrac{A}{180} + \dfrac{B}{180} + \dfrac{C}{180} = 1 + 4(T/S)$$

Multiplying by 180 gives the desired result.

One can easily check that the formula works in the example above with three right-angles, because the triangle in this case covers a quarter of the northern hemisphere, and hence an eighth of the sphere. Also if $T \to 0$ the sum $\to 180º$ as in flat triangles.

A nice corollary to this theorem is a formula giving the sum of the four solid-angles of a tetrahedron in terms of the six edge-angles. I will leave the reader to discover it for herself.

Masterclass Videos

When the DES saw the success of the mathematics masterclasses they suggested making them more widely accessible by putting them on video, and generously offered to provide the necessary funding. I have now completed the first two videos for 13-year-olds, with the help of the producer John Jaworski at the BBC Open University Production Unit, and the DES has given six copies to each Local Education Authority for free copying by any schoolteacher who is interested. The first is on Geometry and Perspective [8], and the second is on Gyroscopes and Boomerangs [9].

I find it takes a long time to make a video, especially if you are trying to put across material that is quite a bit more sophisticated than that which the potential audience is used to. Compared with a live masterclass you have to allow for the fact that you won't be there to answer questions and resolve difficulties. It is a delicate balance between boring them by going too slowly, and losing them by going to fast. It helps to have already practised the material with several classes so as to become familiar with the points which the audience may find difficult. It is necessary to write a very clear detailed script, taking care not to put a foot wrong. It is important to take maximum advantage of the medium so as to present animated graphics, and pictures and experiments that would otherwise not normally be available. At the same time you have to guard against using over-elaborate props to illustrate trivial points, otherwise there is a danger of appearing patronising.

Each video is one hour long, divided into three 20-minute sections, with worksheets to be done between the sections. I found it useful to complete the video before writing the accompanying book containing notes, worksheets, solutions and some slightly harder follow up material. As in the live masterclasses the worksheets are perhaps the most important part of the package since that is where the viewers have a chance to be creative and discover things for themselves. Anything that is taught may soon be forgotten, but anything that you discover for yourself you are unlikely to forget.

The package of video plus book has to be arranged so that it can be used either by a teacher with a class, or by an individual studying alone. Initial assessments of the videos indicate that the top 10% of 13-year-olds can completely understand them working alone, the top 25% can completely understand them with a teacher's help, and the top 60% can get something out of them with the teacher's help.

References

1. M. Emmer, Mathematics and the media: different methodologies, *ICMI Conference on Popularization of Mathematics*, Leeds, 1989.
2. D.A. Quadling, Popularizations of mathematics, *ICMI Conference on Popularization of Mathematics*, Leeds, 1989.
3. S Watson, *Warwick University Mathematics Masterclasses, an evaluation*, MA Thesis 1989, Warwick University.
4. E.C. Zeeman, *Catastrophe theory, selected papers 1972-77*, Addison-Wesley, Reading, Mass., 1977.
5. E.C. Zeeman, *Mathematics into pictures, Royal Institution Christmas Lectures 1978*. Six one-hour lectures on video obtainable through the Royal Institution, 21 Albemarle Street, London W1X 4BS. American NTSC version obtainable from Professor R.H. Abraham, Department of Mathematics, University of California, Santa Cruz, CA 95064, USA.
6. E.C. Zeeman, Population dynamics from game theory, *Global theory of dynamical systems*, Springer Lecture Notes in Mathematics 819 (1980) 471-497.
7. E.C. Zeeman, Dynamics of the evolution of animal conflicts, *J.Theor.Biology* 89 (1981) 247-270.
8. E.C. Zeeman, *Geometry and perspective, Royal Institution Mathematics Masterclass*, 1987 (video and book). Obtainable from the Royal Institution (address above). Loanable from local education authorities in the UK.
9. E.C. Zeeman, *Gyroscopes and boomerangs, Royal Institution Mathematics Masterclass*, 1989 (video and book). Obtainable from the Royal Institution (address above). Loanable from local education authorities in the UK.
10. M.L. Zeeman, Hopf bifurcations in competitive three-dimensional Lotka-Volterra systems, IMA Preprint 622 (1990) University of Minnesota, Minneapolis, USA.

Some Aspects of the Popularization of Mathematics in China

D.Z. ZHANG, H.K. LIU, S. YU

East China Normal University, Shanghai, CHINA

Chinese mathematics has its own age-old tradition. Western mathematics was taught in schools only after the 1910's. Now there are around two million mathematics teachers and mathematicians in China, who are also in a way taking the responsibility for popularizing mathematics. It is probably the largest group of mathematics teachers and mathematicians in any single country in the world.

1. PUBLIC HONOURS

Some Chinese mathematicians by winning public honours have helped to make mathematics more popular in China. The following are two examples.

Hua Loo-Keng (1910-1985) is a household name in China. Colleagues in the mathematics community know the name through Hua's works on number theory. However most of the people in China heard of him because of his relentless effort in popularizing mathematics. Starting from the mid 60's, and apart from continuing his theoretical research, he was actively involved in the popularization of some operational research methods to the general public in China. He and his team visited thousands of factories and farms and met with millions of people over a period of ten to twenty years. As a result, "the optimization method" became a common word in China. Professor Hua reported on some of his experiences in popularizing mathematics in China at ICME 4 in Berkeley, 1980.

The legendary story of how Hua started as a village shop assistant and later became a professor in mathematics, was made into a film script in 1984. Then a television programme on "Hua Loo-Keng" was produced. So far Hua is the only Chinese scientist who has become a leading character in a film during his life time.

Professor Chen Jing-Ren, a student of Hua, is another well-known mathematician in China. He also specializes in number theory, in particular, the Goldbach conjecture. That is, it is conjectured that a large even number must be the sum of two odd prime numbers. In 1973, Chen made an important step toward solving this conjecture. More precisely, he proved that for a given large even number N we could find odd prime numbers P', P'' or P_1, P_2, P_3 such that $N = P' + P''$ or $N = P_1 + P_2 P_3$. Five years later, Xu Chi, a well-known Chinese writer, published a report on the Goldbach conjecture. He described the conjecture as a problem "1 + 1" and Chen's result as "1 + 1" or "1 + 2". Xu depicted vividly how Chen approached the problem, known as the crown jewel of number theory. This report was later reprinted in all the newspapers throughout the country. Even up to now Chen is still regarded as an emblem of hard work. Many gifted students follow Chen's example and study mathematics diligently. So much so that places in the department of mathematics became in great demand at the time and many students studied the subject.

Mathematicians are regarded as a symbol of intelligence in China. Believe it or not, most of the presidents of the well-known universities in China are mathematicians. It is certainly a marvel that this is so.

Some mathematicians are also political figures. A well-known senior geometer, Professor Su Bu-Qing, was elected to be vice-chairman of the Chinese People's Political Consultative Council. Furthermore, at least twenty mathematicians are deputies to the national people's congress at the moment.

2. MATHEMATICAL COMPETITIONS

During the past ten years, the public has paid increasingly greater attention to mathematical competitions. It is a small "hot spot" of social life in China.

The first mathematical competition was held in China in 1956. At present, there are various competitions organized at the levels of country, city and province. The three nation-wide competitions are:

    Hua Loo-Keng Golden Cup competition (age 11-13)
    Junior school mathematical competition (age 13-15)
    Senior school mathematical competition (age 16-18).

About 30,000 pupils take part in each of the above-mentioned competitions. The first 50 participants with the highest marks in each competition will be given awards by the Chinese Mathematical Society. Among the group of age 16-18, 20 students are selected to join a winter training camp for the IMO (International Mathematical Olympiad). After the training camp, six of them are finally selected to form the Chinese IMO team.

The following table shows the results of the Chinese contestants.

1986			1987			1988		
Gold	Silver	Bronze	Gold	Silver	Bronze	Gold	Silver	Bronze
3	1	1	2	2	2	2	4	0

The 17 winners include 3 girls. Like a torch, every winner sparks off the flames of mathematics in his or her home town.

Another exciting event is a competition between children of age 7-8 and the bank clerks. The questions asked are those involving addition, subtraction and multiplication of two-digit numbers. The children using F.S. Shi's fast counting method answer the questions more quickly than the clerks with the aid of calculators. Mr. F.S. Shi has given a performance on his method of calculation at the office of UNESCO. It was said that all the audience applauded the performance.

3. PUBLISHING

Publishing is still the most important means of circulating mathematics in China. We have not really produced any mathematics films or videos. The television programmes on mathematics so far available present only the classroom scene on the screen. More than 1000 books in mathematics are published every year. It is a pity that only those books connected with entrance examinations for high schools and universities might be sold out. A great deal of mathematical publishing incurs a deficit. Some extra-curricular readings are unmarketable.

Recently, some far sighted publishers produced quite a few excellent mathematics series. One of them is "Modern series in science and technology", in which we find booklets with such titles as "Secret codes", "Networks", "Computer graphics", "Distance", "Database systems".

Another series concerns the appreciation of famous mathematics problems. There are 14 books included in the series, for example, "Gödel's theorem", "The Goldbach conjecture", "The Riemann hypothesis", "Fermat's last theorem", "The fixed point theorems", "Fibonacci numbers", "The travelling salesman problem", "Kirkman's girls problem", "Hilbert's tenth problem".

The conditions for the smooth development of the popularization of mathematics in China are not favourable. The size of the population has given us many serious troubles. Since 1985, students of high ability have gradually lost interest in mathematics. In view of this, a master plan of how to popularize mathematics is presently under discussion. The Chinese NSF (National Science Federation) would support the plan.